生物学专业英语教程

主　编　姚晓芹　刘存歧
副主编　楚建周　张艳芬

科学出版社
北京

内 容 简 介

本书共分为三大部分，第一部分是生物学专业英语基础阅读，选编的 19 篇文章涵盖了生物学的主要分支学科的基本内容，通过这部分的学习，学生能够掌握该领域的基本词汇和写作方法。第二部分是生物学专业英语提升篇，选编的 30 篇文章涉及生物学主要分支学科的研究方向和研究前沿，这部分不仅有助于提高学生的阅读能力，而且能够使学生了解生物学主要分支学科的研究内容。第三部分是英语科技论文写作与发表，作者根据一篇具体的论文详细介绍英语科技论文的写作特点，并结合作者多年发表论文中遇到的问题将积累的经验献给读者。

本书可作为高等院校生物学相关专业本科生和研究生的专业英语教材，也可作为相关研究领域科研人员的阅读材料。

图书在版编目（CIP）数据

生物学专业英语教程 / 姚晓芹，刘存歧主编. —北京：科学出版社，2017.6

ISBN 978-7-03-053176-6

Ⅰ. ①生… Ⅱ. ①姚… ②刘… Ⅲ. ①生物学-英语-教材 Ⅳ. ①Q

中国版本图书馆 CIP 数据核字（2017）第 123686 号

责任编辑：滕 云 胡云志/ 责任校对：彭 涛
责任印制：张 伟/ 封面设计：华路天然工作室

科学出版社 出版
北京东黄城根北街 16 号
邮政编码：100717
http: //www. sciencep. com

北京虎彩文化传播有限公司 印刷

科学出版社发行 各地新华书店经销

*

2017 年 6 月第 一 版 开本：720×1000 1/16
2023 年 8 月第六次印刷 印张：12 1/4
字数：290 000

定价：39. 80 元

（如有印装质量问题，我社负责调换）

前　　言

迄今国内大多数高校及大多数专业都开设了专业英语课程，作为21世纪最具发展潜力与前景的生物学专业，其相关领域的最新研究成果多以英文形式发表于本学科主流期刊。因而，生物学相关专业的学生对专业英语的学习就显得更为重要。本书在编写过程中广泛征求学生意见，并基于编者多年生物学专业英语教学实践及多篇英语科技论文撰写和发表经验编写而成，以期能够更好地满足学生对生物学专业英语的需求。

本书主要分三大部分，第一部分是生物学专业英语基础阅读和基本专业词汇。所选文章基本上涵盖细胞、微生物、植物、动物、生态和生物技术的主要内容。通过这部分的学习读者能够掌握相关领域的基本专业词汇和写作特点。第二部分是生物学专业英语提升篇，这一部分所收录的文章都是从高水平的英文期刊上选编而来，并根据需要做了一定的整理和改编。所选论文基本涵盖生物学主要分支学科研究内容和前沿。第三部分是英语科技论文写作与发表。作者以一篇具体论文为例详细介绍了英语科技论文每一部分的写作特点和注意事项，并将作者多年发表英语科技论文所积累的经验一并献给读者。这三部分内容旨在使读者在有限的时间内既能掌握一定量的生物学基本专业词汇，又能具备一定的英语科技论文阅读、翻译和写作能力。

在本书编写过程中，河北大学 2012 级生物科学专业梁娜同学、2013 级生物技术专业范江玲同学，以及 2016 级生态学专业研究生陈国亮、郭春延和张楠同学给予了热情的帮助并提出了很好的建议。另外，本书选编了部分同行作者的文稿内容，在此一并向他们表示诚挚的谢意。

本书是“白洋淀流域生态保护与京津冀可持续发展协同创新”系

列成果之一，感谢其资金的支持。同时，也感谢科学出版社在本书编写和出版方面给予的大力支持和帮助。

本书初稿已先后在河北大学2013级生物科学专业（本科）和2016级生态学专业（研究生）的学生中使用，已将教学过程中发现的问题和错误进行了反复的修订，但可能仍存在不足之处，恳请广大读者不吝赐教（yaoxiao301@126.com），以便再版时做得更好。

姚晓芹

2017年2月

目　　录

Contents

Part I　Base Components
基础篇

Unit 1　Cell …… **3**

Lesson 1　Cell structure …… 3
Lesson 2　Chemical composition of cells …… 7
Lesson 3　Structure and function of DNA …… 11

Unit 2　Microorganism …… **15**

Lesson 4　Types of microorganisms …… 15
Lesson 5　Applications of microorganisms …… 19
Lesson 6　Evolution and morphology of viruses …… 21

Unit 3　Plant …… **25**

Lesson 7　Plant evolution …… 25
Lesson 8　Primary and secondary growth in plants …… 28
Lesson 9　Plant life cycles …… 30
Lesson 10　Natural and artificial methods of asexual reproduction in plants …… 33

Unit 4　Animal …… **36**

Lesson 11　Characteristics of the animal kingdom …… 36
Lesson 12　Methods of animal reproduction …… 39
Lesson 13　Constructing an animal phylogenetic tree …… 41

Unit 5　Ecology …… **45**

Lesson 14　Food chains and food webs …… 45
Lesson 15　Ecosystem dynamics …… 47

Lesson 16 Climate change and biodiversity ······ 49

Unit 6 Biotechnology ······ 52

Lesson 17 Basic techniques to manipulate genetic material ······ 52
Lesson 18 Genetically modified organisms ······ 55
Lesson 19 Basic techniques in protein analysis ······ 58

Part II Promotion Components
提升篇

Unit 7 Cell studies ······ 63

Lesson 20 Mechanisms of plant cell division ······ 63
Lesson 21 Browning phenomena in plant cell cultures ······ 64
Lesson 22 Cell mechanics ······ 66
Lesson 23 Stem cell research: trends and perspectives on the evolving international landscape ······ 67
Lesson 24 Suicide of aging cells prolongs life span in mice ······ 70
Lesson 25 Nobel honors discoveries on how cells eat themselves ······ 72

Unit 8 Microorganism studies ······ 75

Lesson 26 Human adaptation to arsenic-rich environments ······ 75
Lesson 27 Your poor diet might hurt your grandchildren's guts ······ 76
Lesson 28 Antibiotic use and its consequences for the normal microbiome ··· 78
Lesson 29 The global ocean microbiome ······ 81
Lesson 30 Gut microbes give anticancer treatments a boost ······ 83
Lesson 31 Zika virus kills developing brain cells ······ 86

Unit 9 Plant studies ······ 89

Lesson 32 Abiotic and biotic stress combinations ······ 89
Lesson 33 Does climate directly influence NPP globally? ······ 91
Lesson 34 The effects of enhanced UV-B radiation on plants ······ 93
Lesson 35 Sulfur deficiency–induced repressor proteins optimize glucosinolate biosynthesis in plants ······ 94

Lesson 36 Fine roots—functional definition expanded to crop species ······· 97
Lesson 37 Plants can gamble, according to study ······· 100

Unit 10 Animals studies ······· 103

Lesson 38 Marine defaunation: animal loss in the global ocean ······· 103
Lesson 39 Tiny DNA tweaks made snakes legless ······· 104
Lesson 40 How Earth's oldest animals were fossilized ······· 107
Lesson 41 Why do zebras have stripes? ······· 109
Lesson 42 Tiny microbe turns tropical butterfly into male killer ······· 111
Lesson 43 Dogs recognize dog and human emotions ······· 113

Unit 11 Ecology studies ······· 115

Lesson 44 The next century of ecology ······· 115
Lesson 45 Ninety-nine percent of the ocean's plastic is missing ······· 117
Lesson 46 Earth's lakes are warming faster than its air ······· 118
Lesson 47 Strong invaders are strong defenders—implications for the resistance of invaded communities ······· 120
Lesson 48 Global warming favours light-coloured insects in Europe ······· 122
Lesson 49 Carbon dioxide supersaturation promotes primary production in lakes ······· 123

Part III Scientific Writing and Publishing in English
英语科技论文写作与发表

Unit 12 Common problems and corresponding suggestions in document indexing ······· 127

Lesson 50 Common problems in document indexing ······· 127

Unit 13 Composition and writing of English scientific papers ······· 133

Lesson 51 Composition of English scientific papers ······· 133
Lesson 52 Title, author and contact information of the paper ······· 134
Lesson 53 The writing skills of abstract and keywords ······· 136
Lesson 54 The writing skills of the introduction ······· 140

Lesson 55 The writing skills of materials and methods ······ 143
Lesson 56 The writing skills of results, discussions and conclusions ······ 148
Lesson 57 The writing skills of acknowledgements and references ······ 152

Unit 14 Contributions and publishing skills of English scientific papers ······ 156

Lesson 58 How to select the periodicals ······ 156
Lesson 59 Submission process and modification process ······ 159

References ······ 167

Appendix ······ 170

Appendix Ⅰ Common biology core vocabulary ······ 170
Appendix Ⅱ Research paper example ······ 176

Part I　Base Components

基础篇

Biology is one of the six basic natural science disciplines, which is concerned with the study of life and living organisms, including their structure, function, growth, development, and the relationship between biology and environments. Modern biology is a vast field, composed of many branches and subdisciplines. Subdisciplines of biology are defined by the scale at which organisms are studied, the kinds of organisms studied, and the methods used to study them: biochemistry examines the rudimentary chemistry of life; molecular biology studies the complex interactions among biological molecules; botany studies the biology of plants; cellular biology examines the basic building-block of all life,the cell; physiology examines the physical and chemical functions of tissues, organs, and organ systems of an organism; evolutionary biology examines the processes that produce the diversity of life; and ecology examines how organisms interact in their environment. This part mainly summarizes the basic knowledge of the main subdisciplines which are the main courses in the College of Life Science of all university, including cell biology, microbiology, botany, zoology and ecology.

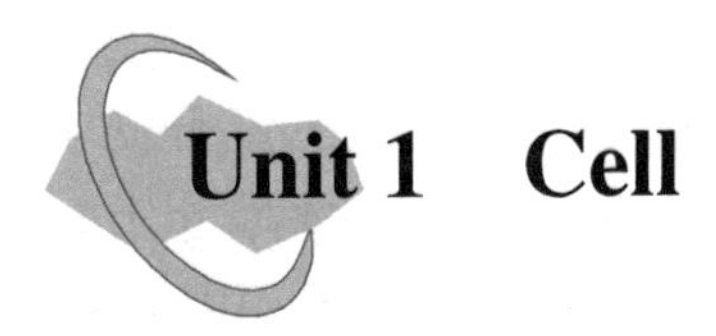

Unit 1 Cell

Cells are the basic units of living organisms. All living things are composed of one or more cells. Although there are many different kinds of cells, they have much in common in structure and chemical composition. This unit mainly introduces the structure and chemical composition of cells, and the structure and function of DNA.

Lesson 1 Cell structure

Most cells are too small to be seen with naked eyes. We can only see them by a microscope. Cells come in two basic types—prokaryotic cells and eukaryotic cells. Prokaryotic cells found in bacteria and the blue-green algae are no membrane bound organelles, no nucleus (only a nuclear region), and they are typically unicellular organisms. Eukaryotic cells are more highly evolved. They have membrane bound organelles (including but not limited to the mitochondria, chloroplast, Golgi apparatus, smooth and rough endoplasmic reticulum), and have a membrane bound nucleus too. Eukaryotic cells are found in animals, plants, fungi and protists. Next, cell structures of eukaryotes are described in detail.

Cell membranes

All living cells have cell membranes (also known as plasmalemma or plasma membrane) composed of phospholipid bilayer with embedded proteins. Cell membrane is a very thin biological membrane that separates the interior of all cells from the outside environment, and is selectively permeable to ions and organic molecules, which control the passage of substances into and out of the cell.

Cell walls

Fungi, bacteria and nearly all plant cells also have cell walls outside of the plasma membrane, which are very different from the plasma membrane. Cell walls can't perform the diffusion barrier tasks of the plasma membrane. One of the primary functions of the cell wall is physical support. Some kinds of bacterial cell walls also have other functions. Prokaryotic cell walls are composed at least partially of an interesting substance called peptidoglycan, which is a kind of hybrid between polysaccharide and protein.

Cytoplasm

The cytoplasm is composed of cytosol and organelles. Within the cells of eukaryotic organisms, the contents of the cell nucleus are separated from the cytoplasm, and are then called the nucleoplasm. The organelle is a specialized subunit within a cell that has a specific function. Individual organelles are usually separately enclosed within their own lipid bilayers. There are many types of organelles in eukaryotic cells.

Cell nucleus

The nucleus of cell is enclosed by a double layer of membrane called nuclear envelope whose function is to confine the materials necessary for DNA and RNA synthesis inside the nucleus, and control the movement into and out of the nucleus. The nucleus contains several—generally two to four—dense structures called nucleoli (singular "nucleolus"). Assembly of ribosomes takes place in nucleoli. The nucleus of a eukaryotic cell contains a number of chromosomes, which are composed of DNA and histone proteins.

Vacuole

Plant cells contain a specialized vacuole called the central vacuole,

which is a large, membrane-bound structure filling in most of the interior of the cell. The central vacuole is filled mostly with water, but always with some impurities—mineral or protein—so the water concentration is always less than 100%. When the cell is surrounded by sufficient water, osmosis makes the central vacuole to swell, and thus makes the cell to press against the inside of the cell wall. This phenomenon in all of the cells of a leaf makes the leaf's tissues to be stiff, and keeps this delicate structure spread out so it can serve its vital function as a solar panel.

Endoplasmic Reticulum

Endoplasmic Reticulum (ER) is a system of membrane-enclosed channels which ramifies throughout the cytoplasm of the cell. It comes in two types—smooth ER and rough ER. The difference is that rough ER has ribosomes all over its outer surface, which performs protein synthesis.

Mitochondria

Mitochondria are very complex, double-membrane-bound organelles. Their function is to perform the aerobic portions of aerobic cellular respiration, the essential energy-producing process of the cell. This is the same function performed by the mesosomes in many prokaryotic cells. Mitochondria contain their own naked, circular DNA and their own ribosomes.

Golgi bodies

Each cell contains a number of Golgi bodies. "Golgi" is the name of the person who first described these structures. Their function is to process materials manufactured by the cell, and then package those products into small structures called "Golgi vesicles". The materials arrive at the Golgi bodies from the smooth endoplasmic reticulum. Golgi vesicles come in two general types—microbodies and secretory vesicles. Microbodies are fated to remain in the cell. They contain materials, usually enzymes, which the cell needs, but which must remain packaged away from the cell's other

contents. The best known of these microbodies is the lysosome. Lysosomes contain digestive enzymes, which, if released into the cell, would digest the vital components of the cell and kill it. "Break" it, in other words.

Plastids

Plant cells contain a family of organelles called plastids. There are several kinds of plastids, all related to each other and, under appropriate conditions, capable of modifying from one type to another. The best known of these plastids is the chloroplast, which performs the function of photosynthesis. Chloroplasts are double-membrane-bound, like mitochondria. Also like mitochondria, their inner membrane is very complicated. In fact, it's formed into many thylakoid structures which perform the same function the thylakoids done in prokaryotic cells.

Glossary

microscope 显微镜
prokaryotic 原核的
eukaryotic 真核的
bacterium 细菌
blue-green algae 蓝绿藻
fungi 真菌
protist 原生生物
eukaryote 真核生物
cell membrane 细胞膜
phospholipid 磷脂
protein 蛋白
cell wall 细胞壁
peptidoglycan 肽聚糖
polysaccharide 多糖
cytoplasm 细胞质
cytosol 细胞液
organelle 细胞器
nucleoplasm 核质
subunit 亚单位
lipid bilayer 脂类双分子层
nucleus 细胞核
nuclear envelope 核膜
nucleoli 核仁
ribosome 核糖体
chromosome 染色体
histone protein 组蛋白
vacuole 液泡
central vacuole 中央液泡
osmosis 渗透
Endoplasmic Reticulum (ER) 内质网
smooth ER 光滑内质网
rough ER 粗糙内质网
mitochondrion 线粒体
mesosome 中间体
Golgi body 高尔基体
microbody 微体

secretory 分泌的、分泌腺
enzyme 酶
lysosome 溶酶体
plastid 质体
chloroplast 叶绿体
photosynthesis 光合作用
thylakoid 类囊体

Lesson 2 Chemical composition of cells

All living organisms, from microbes to mammals, are composed of chemical substances from both the inorganic and organic world, which appear in roughly the same proportions, and perform the same general tasks.

The basic chemical elements of cells include hydrogen, oxygen, nitrogen, carbon, phosphorus and sulfur, which normally make up more than 99% of the mass of living cells, and when combined in various ways, form virtually all known organic biomolecules. There are four general classes of macromolecules within living cells: nucleic acids, proteins, polysaccharides and lipids.

Nucleic acids

Nucleic acids are nucleotide polymer that store and transmit genetic information. Only 4 different nucleotides are used in nucleic acid biosynthesis. Genetic information contained in nucleic acids is stored and replicated in chromosomes, which contain genes. A chromosome is a deoxyribonucleic acid (DNA) molecule, and genes are segments of intact DNA. When a cell replicates itself, identical copies of DNA molecules are produced, therefore the hereditary line of descent is conserved, and the genetic information carried on DNA is available to direct the occurrence of virtually all chemical reactions within the cell. The bulk of genetic information carried on DNA provides instructions for the assembly of virtually every protein molecule within the cell.

Proteins

Proteins are amino acid polymers responsible for implementing instructions contained within the genetic code. There are twenty of amino acids used to synthesize proteins. A typical cell contains thousands of different proteins that perform a vast array of functions within organisms, including catalysing metabolic reactions, DNA replication, responding to stimuli, and transporting molecules from one location to another. Proteins differ from one another primarily in their sequence of amino acids, which is dictated by the nucleotide sequence of their genes, and which usually results in protein folding into a specific three-dimensional structure that determines its activity.

Most proteins fold into unique 3-dimensional structures. The shape into which a protein naturally folds is known as its native conformation. Although many proteins can fold unassisted, simply through the chemical properties of their amino acids, others require the aid of molecular chaperones to fold into their native states. Biochemists often refer to four distinct aspects of a protein's structure: Primary structure refers to the amino acid sequence. Secondary structure is formed by regularly repeating local structures stabilized by hydrogen bonds. The most common examples are the α-helix, β-sheet and turns. Because secondary structures are local, many regions of different secondary structure can be present in the same protein molecule. Protein tertiary structure refers to a protein's geometric shape. The tertiary structure will have a single polypeptide chain "backbone" with one or more protein secondary structures, the protein domains. Amino acid side chains may interact and bond in a number of ways. The interactions and bonds of side chains within a particular protein determine its tertiary structure. A number of tertiary structures may fold into a quaternary structure.

Proteins can be informally divided into three main classes, which correlate with typical tertiary structures: globular proteins, fibrous proteins, and membrane proteins. Almost all globular proteins are soluble and many are enzymes. Fibrous proteins are often structural, such as collagen, the major component of connective tissue, or keratin, the protein component of hair and nails. Membrane proteins often serve as receptors or provide

channels for polar or charged molecules to pass through the cell membrane.

Polysaccharides

Polysaccharides are polymers of simple sugars (i.e., monosaccharides). Some polysaccharides are homogeneous polymers that contain only one kind of sugar (e.g., glycogen), while others are complex heterogenous polymers that contain 8-10 types of sugars. Polysaccharides can occur as functional and structural components of cells (e.g., cellulose and chitin), or merely as storage forms of energy (e.g., glycogen and starch). The 8-10 monosaccharides that become the building blocks for heterogenous polysaccharides can be synthesized from glucose, or formed from other metabolic intermediates.

Nutrition polysaccharides are common sources of energy. Many organisms can easily break down starches into glucose; however, most organisms cannot metabolize cellulose or other polysaccharides like chitin and arabinoxylans. These carbohydrate types can be metabolized by some bacteria and protists. Ruminants and termites, for example, use microorganisms to process cellulose.

Even though these complex carbohydrates are not very digestible, they provide important dietary elements for humans. Called dietary fiber, these carbohydrates enhance digestion among other benefits. The main action of dietary fiber is to change the nature of the contents of the gastrointestinal tract, and to change how other nutrients and chemicals are absorbed. Soluble fiber binds to bile acids in the small intestine, making them less likely to enter the body; this in turn lowers cholesterol levels in the blood. Soluble fiber also attenuates the absorption of sugar, reduces sugar response after eating, normalizes blood lipid levels and, once fermented in the colon, produces short-chain fatty acids as byproducts with wide-ranging physiological activities (discussion below). Although insoluble fiber is associated with reduced diabetes risk, the mechanism by which this occurs is unknown.

Not yet formally proposed as an essential macronutrient, dietary fiber is nevertheless regarded as important for the diet, with regulatory authorities in many developed countries recommending increases in fiber intake.

Lipids

Lipids are nonpolar substances that are mostly insoluble in water, yet soluble in nonpolar solvents (like chloroform), which comprise a group of naturally occurring molecules that include fats, waxes, sterols, fat-soluble vitamins (such as vitamins A, D, E, and K), monoglycerides, diglycerides, triglycerides, phospholipids, and others. The main biological functions of lipids serve as membrane components (cholesterol, glycolipids and phospholipids), storage forms of energy (triglycerides), precursors to other important biomolecules (fatty acids), protective coatings to prevent infection and excessive gain or loss of water, and some vitamins (A, D, E, and K) and hormones (steroid hormones).

Major classes of lipids are the saturated and unsaturated fatty acids. All lipids can be synthesized from acetyl-CoA, which in turn can be generated from numerous different sources, including carbohydrates, amino acids, short-chain volatile fatty acids (e.g., acetate), ketone bodies, and fatty acids.

The term lipid is sometimes used as a synonym for fats, but, fats are a subgroup of lipids called triglycerides. Humans and other mammals use various biosynthetic pathways both to break down and to synthesize lipids. However, some essential lipids cannot be made by this way and must be obtained from the diet.

mammal 哺乳动物
inorganic 无机的
organic 有机的
biomolecule 生物分子
macromolecule 大分子
nucleic acid 核酸
nucleotide 核苷酸
biosynthesis 生物合成
chromosome 染色体
deoxyribonucleic acid 脱氧核糖核酸
amino acid 氨基酸
tertiary 三级的
polysaccharide 多糖
monosaccharide 单糖
heterogenous 异源的
cellulose 纤维素

chitin 几丁质
lipids 脂类
nonpolar solvent 非极性溶剂
cholesterol 胆固醇
fatty acid 脂肪酸
hormone 激素
acetyl-CoA 乙酰辅酶 A
carbohydrate 碳水化合物
acetate 乙酸盐、乙酸酯
ketone body 酮体

Lesson 3 Structure and function of DNA

Deoxyribonucleic acid (DNA) is a molecule that carries most of the genetic instructions used in the development, functioning and reproduction of all known living organisms and many viruses.

DNA is a nucleic acid; alongside proteins and carbohydrates, nucleic acids are one of the three major macromolecules essential for all known forms of life. DNA stores biological information and is involved in the <u>expression</u> of traits in all living organisms.

In the 1950s, Francis Crick and James Watson worked together to determine the structure of DNA at the University of Cambridge, England. At the time, other scientists like Linus Pauling and Maurice Wilkins were also actively exploring this field. Pauling had discovered the secondary structure of proteins using <u>X-ray crystallography</u>.

The structure of DNA

The monomeric building blocks of DNA are deoxyribomono-nucleotides (usually referred to as just nucleotides), and DNA is formed from linear chains, or polymers, of these nucleotides. The components of the nucleotide used in DNA synthesis are a nitrogenous base, a deoxyribose, and a phosphate group. The nucleotide is named depending on which nitrogenous base is present. The nitrogenous base can be a purine such as <u>adenine</u> (A) and <u>guanine</u> (G), characterized by double-ring structures, or a pyrimidine such as <u>cytosine</u> (C) and <u>thymine</u> (T), characterized by single-ring structures. In polynucleotides (the linear polymers of nucleotides) the nucleotides are connected to each other by covalent bonds known as phosphodiester bonds or phosphodiester linkages.

Basics of DNA replication

Watson and Crick's discovery that DNA is a two-stranded double helix provided a hint as to how DNA is replicated. During cell division, each DNA molecule has to be perfectly copied to ensure identical DNA molecules to move to each of the two daughter cells. The double-stranded structure of DNA suggested that the two strands might separate during replication with each strand serving as a template from which the new complementary strand for each is copied, generating two double-stranded molecules from one.

Models of replication

There were three models of replication possible from such a scheme: conservative, semi-conservative, and dispersive. In conservative replication, the two original DNA strands, known as the parental strands, would re-basepair with each other after being used as templates to synthesize new strands; and the two newly-synthesized strands, known as the daughter strands, would also basepair with each other; one of the two DNA molecules after replication would be "all-old" and the other would be "all-new". In semi-conservative replication, each of the two parental DNA strands would act as a template for new DNA strands to be synthesized, but after replication, each parental DNA strand would basepair with the complementary newly-synthesized strand just synthesized, and both double-stranded DNAs would include one parental or "old" strand and one daughter or "new" strand. In dispersive replication, after replication both copies of the new DNAs would somehow have alternating segments of parental DNA and newly-synthesized DNA on each of their two strands.

DNA repair

DNA replication is a highly accurate process, but mistakes can

occasionally occur when a DNA polymerase inserts a wrong base. Uncorrected mistakes may sometimes lead to serious consequences, such as cancer. Repair mechanisms can correct the mistakes, but in rare cases mistakes are not corrected, leading to mutations; in other cases, repair enzymes are themselves mutated or defective.

Most of the mistakes during DNA replication are promptly corrected by DNA polymerase which proofreads the base that has just been added. In proofreading, the DNA pol reads the newly-added base before adding the next one so a correction can be made. The polymerase checks whether the newly-added base has paired correctly with the base in the template strand. If it is the correct base, the next nucleotide is added. If an incorrect base has been added, the enzyme makes a cut at the phosphodiester bond and releases the incorrect nucleotide. This is performed by the exonuclease action of DNA pol Ⅲ. Once the incorrect nucleotide has been removed, a new one will be added again.

Some errors are not corrected during replication, but are instead corrected after replication is completed; this type of repair is known as mismatch repair. The enzymes recognize the incorrectly-added nucleotide and excise it; this is then replaced by the correct base. If this remains uncorrected, it may lead to more permanent damage. How do mismatch repair enzymes recognize which of the two bases is the incorrect one? In *E. coli*, after replication, the nitrogenous base adenine acquires a methyl group; the parental DNA strand will have methyl groups, whereas the newly-synthesized strand lacks them. Thus, DNA polymerase is able to remove the incorrectly-incorporated bases from the newly-synthesized, non-methylated strand. In eukaryotes, the mechanism is not very well understood, but it is believed to involve recognition of unsealed nicks in the new strand, as well as a short-term continuing association of some of the replication proteins with the new daughter strand after replication has been completed.

In another type of repair mechanism, nucleotide excision repair, enzymes replace incorrect bases by making a cut on both the 3′ and 5′ ends of the incorrect base. The segment of DNA is removed and replaced with the correctly-paired nucleotides by the action of DNA pol. Once the

bases are filled in, the remaining gap is sealed with a phosphodiester linkage catalyzed by DNA ligase. This repair mechanism is often employed when UV exposure causes the formation of pyrimidine dimers.

Glossary

expression 表达
X-ray crystallography X 射线晶体学
adenine 腺嘌呤
guanine 鸟嘌呤
cytosine 胞嘧啶
thymine 胸腺嘧啶
phosphodiester 磷酸二酯
basepair 碱基对
DNA polymerase DNA 聚合酶
proofread 校对
exonuclease 核酸外切酶
DNA ligase DNA 连接酶

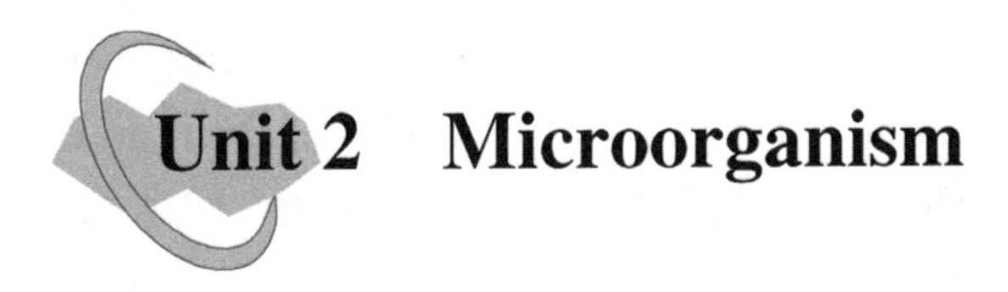

Unit 2 Microorganism

A microorganism is a microscopic living organism, which may be single celled or multicellular. The study of microorganisms is called microbiology, a subject that begins with the discovery of microorganisms in 1674 by Antonie van Leeuwenhoek, using a microscope of his own design. Viruses are generally regarded as not living and therefore strictly speaking are not microbes, although the field of microbiology also encompasses the study of viruses. This unit mainly introduces the types and applications of microorganisms, and evolution and morphology of viruses.

Lesson 4 Types of microorganisms

Microorganisms can be found almost anywhere in the taxonomic organization of life on the planet. Bacteria and archaea are almost always microscopic, while a number of eukaryotes are also microscopic, including most protists and a number of fungi. Next, different types of microorganisms will be introduced one by one.

Bacteria

The prokaryotic bacteria are the simplest and the most diverse and widespread group of organisms on Earth. Bacteria inhabit practically all environments where some liquid water is available and the temperature is below 140°C. They are found in sea water, soil, the gastrointestinal tract, hot springs, and in food. Practically all surfaces that have not been specially sterilized are covered in bacteria. The number of bacteria in the world is estimated to be around five million trillion, or 5×10^{30}.

Bacteria are practically all invisible to the naked eye, with few extremely rare exceptions, such as *Thiomargarita namibiensis*. They are unicellular organisms and lack organelles, including a nucleus. Their genome is usually a single string of DNA, although some of them harbor small pieces of DNA called plasmids. Bacteria are surrounded by a cell wall. They reproduce asexually by binary fission. Some species form spores, but for bacteria this is a mechanism for survival, not reproduction. Under optimal conditions, bacteria can grow extremely rapidly and have been reported as doubling as quickly as every ten minutes.

Archaea

Archaea are also single-celled organisms that lack nuclei. Archaea differ from bacteria in both their genetics and biochemistry. For example, while bacterial cell membranes are made from phosphoglycerides with ester bonds, archaean membranes are made of ether lipids. These organisms are also common in soil and play a vital role in ammonia oxidation.

Protists

Of eukaryotic groups, the protists are most commonly unicellular. This is a highly diverse group of organisms that are not easy to classify.

Several algae species are multicellular protists. The number of species of protists is unknown since we may have identified only a small portion. Studies from 2001 to 2004 have shown that a high degree of protist diversity exists in oceans, deep sea-vents, river sediment and an acidic river which suggests that a large number of eukaryotic microbial communities have yet to be discovered.

Fungi

Fungi are any member of the group of eukaryotic organisms that includes unicellular microorganisms such as yeasts, as well as multicellular fungi that produce familiar fruiting forms known as mushrooms. These organisms are classified as a kingdom, Fungi, which separate from the

other life kingdoms of plants, animals, protists, and bacteria. Fungi reproduce both asexually, by budding or binary fission, as well by producing spores, which are called conidia when produced asexually, or basidiospores when produced sexually. Many fungi are extremely useful for human. Many species produce metabolites that are major sources of pharmacologically active drugs. Particularly important are the antibiotics, including the penicillins; many fungi are used in industries, for example, the manufacture of both bread and beer depends on the biochemical activities of yeasts, single-celled fungi that produce abundant quantities of ethanol and carbon dioxide. However, some fungi are harmful because they decay and spoil many different materials as they obtain food.

Green algae

The green algae are a large group of photosynthetic eukaryotes that include many microscopic organisms. There are about 6000 species of green algae. Although some green algae are classified as protists, others such as charophyta are classified with embryophyte plants, which are the most familiar group of land plants. Algae can grow as single cells, or in long chains of cells. The green algae include unicellular and colonial flagellates, usually but not always with two flagella per cell, as well as various colonial, coccoid, and filamentous forms.

Micro animals

Some micro animals are multicellular but at least one animal group, Myxozoa is unicellular in its adult form. Microscopic arthropods include dust mites and spider mites. Microscopic crustaceans include copepods, some cladocera and water bears. Many nematodes are also too small to be seen with the naked eye. Some micro-animals reproduce both sexually and asexually and may reach new habitats by producing eggs which can survive harsh environments that would kill the adult animal. However, some simple animals, such as rotifers, tardigrades and nematodes, can dry out completely and remain dormant for long periods of time.

Glossary

microorganism　微生物
taxonomic　分类的、分类学的
archaea　古生菌
microscopic　微小的
virus　病毒
gastrointestinal tract　胃肠道
Thiomargarita namibiensis　纳米比亚嗜硫珠菌
unicellular　单细胞的
genome　基因组、染色体组
genetics　遗传学
biochemistry　生化
phosphoglyceride　甘油磷脂
ester bond　酯键
ammonia oxidation　氨的氧化
yeast　酵母
mushroom　蘑菇
asexually　无性地
budding　出芽
binary fission　二分体
conidia　分生孢子
basidiospore　担孢子
sexually　性别上地
metabolite　代谢物
antibiotics　抗生素
penicillin　青霉素
photosynthetic　光合的
eukaryote　真核细胞
charophyta　轮藻门
embryophyte　有胚植物
myxozoa　粘原虫门
microscopic arthropod　微型节肢动物
dust mite　尘螨
spider mite　叶螨
microscopic crustacean　微小甲壳纲动物
copepod　桡足类
cladocera　枝角目
water bear　缓步纲动物
nematode　线虫
rotifer　轮虫类
tardigrade　缓步类

Lesson 5 Applications of microorganisms

Microorganisms are vital to humans and the environment, as they participate in the carbon and nitrogen cycles, as well as fulfilling other vital roles in virtually all ecosystems, such as recycling other organisms' dead remains and waste products through decomposition. Microorganisms also play an important role in most higher-order multicellular organisms as symbionts.

Food production

Microorganisms are used in brewing, wine making, baking, pickling and other food-making processes. They are also used to control the fermentation process in the production of cultured dairy products such as yogurt and cheese. The cultures also provide flavor and aroma, and inhibit undesirable organisms.

Water treatment

The majority of all oxidative sewage treatment processes rely on a large range of microorganisms to oxidise organic constituents which are not amenable to sedimentation or flotation. Anaerobic microorganisms are also used to reduce sludge solids producing methane gas (amongst other gases) and a sterile mineralised residue. In potable water treatment, one method, the slow sand filter, employs a complex gelatinous layer composed of a wide range of microorganisms to remove both dissolved and particulate material from raw water.

Energy

Microorganisms are used in fermentation to produce ethanol, and in biogas reactors to produce methane. Scientists are researching the use of algae to produce liquid fuels, and bacteria to convert various forms of agricultural and urban waste into usable fuels.

Chemicals

Microorganisms are used for many commercial and industrial productions of chemicals, enzymes and other bioactive molecules. Examples of organic acid produced include: Acetic acid is produced by the bacterium Acetobacter aceti and other acetic acid bacteria (AAB); Butyric acid (butanoic acid) is produced by the bacterium *Clostridium butyricum*; Citric acid is produced by the fungus *Aspergillus niger*.

Science

Microorganisms are essential tools in biotechnology, biochemistry, genetics, and molecular biology. The yeasts (*Saccharomyces cerevisiae*) and fission yeast (*Schizosaccharomyces pombe*) are important model organisms in science, since they are simple eukaryotes that can be grown rapidly in large numbers and are easily manipulated. They are particularly valuable in genetics, genomics and proteomics. Microorganisms can be harnessed for uses such as creating steroids and treating skin diseases. Scientists are also considering using microorganisms for living fuel cells, and as a solution for pollution.

Human digestion

Microorganisms can form an endosymbiotic relationship with other, larger organisms. For example, the bacteria that live within the human digestive system contribute to gut immunity, synthesise vitamins such as folic acid and biotin, and ferment complex indigestible carbohydrates.

Ecology

Microorganisms play critical roles in Earth's biogeochemical cycles as they are responsible for decomposition and play vital parts in carbon and nitrogen fixation, as well as oxygen production.

Glossary

symbiont 共生体
brewe 酿造
fermentation 发酵
yogurt 酸奶
cheese 奶酪
sewage 污水
sedimentation 沉淀
flotation 漂浮
anaerobic microorganisms 厌氧微生物
sludge 污泥
acetic acid 乙酸
acetobacter aceti 醋酸杆菌
butyric acid 丁酸
Clostridium butyricum 丁酸梭菌
citric acid 柠檬酸
Aspergillus niger 黑曲霉
biotechnology 生物技术
endosymbiotic 内共生的
vitamin 维生素
folic acid 叶酸
biotin 生物素、维生素 H

Lesson 6 Evolution and morphology of viruses

Viruses were first discovered after the development of a porcelain filter, called the Chamberland-Pasteur filter, which could remove all bacteria visible in the microscope from any liquid sample. In 1886, Adolph Meyer demonstrated that a disease of tobacco plants, tobacco mosaic disease, could be transferred from a diseased plant to a healthy one via liquid plant extracts. In 1892, Dmitri Ivanowski showed that this disease could be transmitted in this way even after the Chamberland-Pasteur filter had removed all viable bacteria from the extract. Still, it was many years before it was proven that these "filterable" infectious agents were not simply very small bacteria, but were a new type of tiny, disease-causing particle.

Evolution of viruses

Although biologists have accumulated a significant amount of knowledge about how present-day viruses evolve, much less is known about how viruses originated in the first place. When exploring the evolutionary history of most organisms, scientists can look at fossil records and similar historic evidence. However, viruses do not fossilize, so researchers must conjecture by investigating how today's viruses evolve and by using biochemical and genetic information to create speculative virus histories.

While most findings agree that viruses don't have a single common ancestor, scholars have yet to find one hypothesis about virus origins that is fully accepted in the field. One possible hypothesis, called devolution or the regressive hypothesis, proposes to explain the origin of viruses by suggesting that viruses evolved from free-living cells. However, many components of how this process might have occurred are a mystery. A second hypothesis (called escapist or the progressive hypothesis) accounts for viruses having either an RNA or a DNA genome and suggests that viruses originated from RNA and DNA molecules that escaped from a host cell. A third hypothesis posits a system of self-replication similar to that of other self-replicating molecules, probably evolving alongside the cells they rely on as hosts; studies of some plant pathogens support this hypothesis.

As technology advances, scientists may develop and refine further hypotheses to explain the origin of viruses. The emerging field called virus molecular systematics attempts to do just that through comparisons of sequenced genetic material. These researchers hope to one day better understand the origin of viruses, a discovery that could lead to advances in the treatments for the ailments they produce.

Viral Morphology

Viruses are acellular, meaning they are biological entities that do not have a cellular structure. Therefore, they lack most of the components of cells, such as organelles, ribosomes, and the plasma membrane. A virion consists of a nucleic acid core, an outer protein coating or capsid, and sometimes an outer envelope made of protein and phospholipid membranes derived from the host cell. The capsid is made up of protein subunits called capsomeres. Viruses may also contain additional proteins, such as enzymes. The most obvious difference between members of viral families is their morphology, which is quite diverse. An interesting feature of viral complexity is that host and virion complexity are uncorrelated. Some of the

most intricate virion structures are observed in bacteriophages, viruses that infect the simplest living organisms: bacteria.

Viruses come in many shapes and sizes, but these are consistent and distinct for each viral family. In general, the shapes of viruses are classified into four groups: filamentous, isometric (or icosahedral), enveloped, and head and tail. Filamentous viruses are long and cylindrical. Many plant viruses are filamentous, including TMV (tobacco mosaic virus). Isometric viruses have shapes that are roughly spherical, such as poliovirus or herpesviruses. Enveloped viruses have membranes surrounding capsids. Animal viruses, such as HIV, are frequently enveloped. Head and tail viruses infect bacteria. They have a head that is similar to icosahedral viruses and a tail shape like filamentous viruses.

Many viruses use some sort of glycoprotein to attach to their host cells via molecules on the cell called viral receptors. For these viruses, attachment is a requirement for later penetration of the cell membrane, allowing them to complete their replication inside the cell. The receptors that viruses use are molecules that are normally found on cell surfaces and have their own physiological functions. Viruses have simply evolved to make use of these molecules for their own replication.

Overall, the shape of the virion and the presence or absence of an envelope tell us little about what disease the virus may cause or what species it might infect, but they are still useful means to begin viral classification. Among the most complex virions known, the T4 bacteriophage, which infects the *Escherichia coli* bacterium, has a tail structure that the virus uses to attach to host cells and a head structure that houses its DNA. Adenovirus, a non-enveloped animal virus that causes respiratory illnesses in humans, uses glycoprotein spikes protruding from its capsomeres to attach to host cells. Non-enveloped viruses also include those that cause polio (poliovirus), plantar warts (papillomavirus), and hepatitis A (hepatitis A virus).

Types of nucleic acid

Unlike nearly all living organisms that use DNA as their genetic material, viruses may use either DNA or RNA. The virus core

contains the genome or total genetic content of the virus. Viral genomes tend to be small, containing only those genes that encode proteins that the virus cannot obtain from the host cell. This genetic material may be single- or double-stranded. It may also be linear or circular. While most viruses contain a single nucleic acid.

In DNA viruses, the viral DNA directs the host cell's replication proteins to synthesize new copies of the viral genome and to transcribe and translate that genome into viral proteins. DNA viruses cause human diseases, such as chickenpox, hepatitis B, and some venereal diseases, like herpes and genital warts.

RNA viruses contain only RNA as their genetic material. To replicate their genomes in the host cell, the RNA viruses encode enzymes that can replicate RNA into DNA, which cannot be done by the host cell. These RNA polymerase enzymes are more likely to make copying errors than DNA polymerases and, therefore, often make mistakes during transcription. For this reason, mutations in RNA viruses occur more frequently than in DNA viruses. This causes them to change and adapt more rapidly to their host. Human diseases caused by RNA viruses include hepatitis C, measles, and rabies.

Glossary

evolve 发展，进化
fossil 化石
conjecture 推测
ancestor 祖先
pathogen 病原体
acellular 非细胞的
capsid 衣壳
bacteriophage 细菌噬菌体
filamentous 细丝状的
isometric 等轴的；等比例的
glycoprotein 糖蛋白类
T4 bacteriophage T4 噬菌体
Escherichia coli bacterium 大肠杆菌
poliovirus 脊髓灰质炎病毒
plantar warts 足底疣
hepatitis A 甲型肝炎
chickenpox 水痘
hepatitis B 乙型肝炎

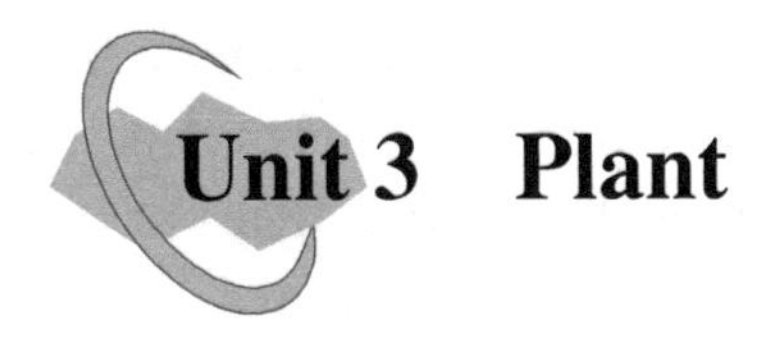

Unit 3 Plant

Plants, also called green plants, are multicellular eukaryotes of the kingdom Plantae. They are essential in nature, and widely distributed on the land, in the rivers, lakes and oceans, and even high above a forest floor. There are about 500,000 species of plants in the world. The evolution of plants from ancestral algae to the more recent taxa has involved a number of important innovations, including adaptations for land colonization (e.g., cuticle, vascularization, stomata), increase in body size (e.g., cell walls with lignin, wood), and more specialized reproduction (e.g., pollen, seeds, double fertilization). Within the plant kingdom the dominance of the two generations varies. Nonvascular plants (Bryophytes) have the gametophyte generation dominant (as do most algae of the protista). Vascular plants (Pteridophytes, Gymnosperms, and Angiosperms) show a progression of increasing sporophytes dominance from the ferns to angiosperms. This unit mainly introduces plant evolution, primary and secondary growth in plants, plant life spans, and natural and artificial methods of asexual reproduction in plants.

Lesson 7 Plant evolution

Plants can be divided into four types based on their evolutional levels: bryophytes, pteridophyta, gymnosperms and angiosperms. Bryophytes belong to nonvascular plants. Meanwhile, pteridophyta, gymnosperms and angiosperms have vascular tissues, so they are also known as vascular plants. Next, let's relate the four types of plants.

Bryophytes

Bryophytes are a group of simple and primitive plants well-adapted to moist habitats. They are basically terrestrial forms, but require the presence of water for completing the life cycle. Hence, they are commonly described as amphibians of the plant kingdom. Their plant height is generally not more than 20 cm. The life cycle of bryophytes shows two distinct phases namely a haploid gametophytic phase and a diploid sporophytic phase alternating with each other. The adult plant body represents the gametophyte. A short-lived sporophyte occurs as a parasite on the gametophyte. Roots are absent and instead thread-like rhizoids are present. Vascular tissues xylem and phloem are absent. Water and food are directly transported from a cell to cell. Bryophytes have two means of reproduction: sexual reproduction and asexual reproduction. The former requires the presence of water; however, the latter can be independent of water.

Pteridophytes

Pteridophytes are a relatively lower division in higher plants, and are a group of primitive vascular plants that include horsetails, clubmosses, spikemosses and quillworts. Pteridophytes differ from mosses and seed plants in that both generations (gametophyte and sporophtye) are independent and free-living, although the sporophyte is generally much larger and more conspicuous. Pteridophytes have the differentiations of root, stem and leaf, but they produce neither flower nor seed. So, they reproduce with spores that germinate only in moist areas, or using rhizomes.

Gymnosperms

The gymnosperms are a group of seed-producing plants that includes conifers, cycads, ginkgoes, and gnetales. Gymnosperms, like all vascular plants, have a sporophyte-dominant life cycle. The gametophyte is relatively short-lived. Two spore types, microspores and megaspores, are typically produced in pollen cones or ovulate cones, respectively. They have no real flowers and still use strobilus as their reproductive organs. The fertilization processes do not depend on water, and two main modes of fertilization are found in gymnosperms. Cycads and Ginkgo have motile

sperm that swim directly to the egg inside the ovule, whereas conifers and gnetophytes have sperm with no flagella that are conveyed to the egg along a pollen tube.

Angiosperms

Angiosperms are flowering plants. Angiosperms are the biggest group in the plant kingdom. They have true roots, stems, leaves and flowers. They also have seeds. The seeds are formed when an egg or ovule is fertilized by pollen in the ovary. The ovary is within a flower. The flower contains the male and/or female parts of the plant. Fruits are frequently produced from these ripened ovaries.

Angiosperms are more highly evolved plant groups. Their advanced structures allow angiosperms to thrive on land. They have roots that hold the plant in place and take in needed minerals and water. They have leaves that are the major food makers for the plant. They have stems that hold the plants up and move the nutrients and water about the plant.

In addition, angiosperms are the primary food source for animals and provide oxygen for us to breathe. They provide lumber for buildings and other objects, fibers for clothes, and the basis for many drugs, etc.

Glossary

bryophyte 苔藓植物
pteridophyta 蕨类植物
gymnosperm 裸子植物
angiosperm 被子植物
nonvascular 非维管的
vascular 维管的
amphibian 两栖动物
haploid 单倍体
diploid 二倍体
gametophyte 配子体
sporophyte 孢子体
rhizoid 假根
xylem 木质部
phloem 韧皮部
sexual reproduction 有性繁殖
asexual reproduction 无性繁殖
horsetail 木贼类
clubmoss 石松类
spikemoss 卷柏类
quillwort 水韭属植物
moss 苔藓植物
spore 孢子
conifer 松柏类
cycad 苏铁植物

ginkgo 银杏
gnetales 买麻藤目
microspore 小孢子
megaspore 大孢子
pollen 花粉
ovulate vi. 排卵，产卵；adj. 具胚珠的
strobilus 孢子叶球
ovary 子房

Lesson 8 Primary and secondary growth in plants

Growth in plants occurs as the stems and roots lengthen. Some plants, especially those that are woody, also increase in thickness during their life span. The increase in length of the shoot and the root is referred to as primary growth. It is the result of cell division in the shoot apical meristem. Secondary growth is characterized by an increase in thickness or girth of the plant. It is caused by cell division in the lateral meristem. Herbaceous plants mostly undergo primary growth, with little secondary growth or increase in thickness. Secondary growth, or "wood", is noticeable in woody plants; it occurs in some dicots, but occurs very rarely in monocots.

Some plant parts, such as stems and roots, continue to grow throughout a plant's life: a phenomenon called indeterminate growth. Other plant parts, such as leaves and flowers, exhibit determinate growth, which ceases when a plant part reaches a particular size. Next, the primary growth and secondary growth in plants are illustrated by the example of stem.

Primary growth

Most primary growth occurs at the apices, or tips, of stems and roots. Primary growth is a result of rapidly-dividing cells in the apical meristems at the shoot tip and root tip. Subsequent cell elongation also contributes to primary growth. The growth of shoots and roots during primary growth enables plants to continuously seek water (roots) or sunlight (shoots).

The influence of the apical bud on overall plant growth is known as apical dominance, which diminishes the growth of axillary buds that form along the sides of branches and stems. Most coniferous trees exhibit strong apical dominance, thus producing the typical conical Christmas tree shape. If the apical bud is removed, then the axillary buds will start forming

lateral branches. Gardeners make use of this fact when they prune plants by cutting off the tops of branches, thus encouraging the axillary buds to grow out, giving the plant a bushy shape.

Secondary growth

The increase in stem thickness that results from secondary growth is due to the activity of the lateral meristems, which are lacking in herbaceous plants. Lateral meristems include the vascular cambium and, in woody plants, the cork cambium. The vascular cambium is located just outside the primary xylem and to the interior of the primary phloem. The cells of the vascular cambium divide and form secondary xylem (tracheids and vessel elements) to the inside and secondary phloem (sieve elements and companion cells) to the outside. The thickening of the stem that occurs in secondary growth is due to the formation of secondary phloem and secondary xylem by the vascular cambium, plus the action of cork cambium, which forms the tough outermost layer of the stem. The cells of the secondary xylem contain lignin, which provides hardiness and strength.

In woody plants, cork cambium is the outermost lateral meristem. It produces cork cells (bark) containing a waxy substance known as suberin that can repel water. The bark protects the plant against physical damage and helps reduce water loss. The cork cambium also produces a layer of cells known as phelloderm, which grows inward from the cambium. The cork cambium, cork cells, and phelloderm are collectively termed the periderm. The periderm substitutes for the epidermis in mature plants. In some plants, the periderm has many openings, known as lenticels, which allow the interior cells to exchange gases with the outside atmosphere. This supplies oxygen to the living- and metabolically-active cells of the cortex, xylem, and phloem.

The activity of the vascular cambium gives rise to annual growth rings. During the spring growing season, cells of the secondary xylem have a large internal diameter; their primary cell walls are not extensively thickened. This is known as early wood, or spring wood. During the fall season, the secondary xylem develops thickened cell walls, forming late wood, or autumn wood, which is denser than early wood. This alternation

of early and late wood is due largely to a seasonal decrease in the number of vessel elements and a seasonal increase in the number of tracheids. It results in the formation of an annual ring, which can be seen as a circular ring in the cross section of the stem. An examination of the number of annual rings and their nature (such as their size and cell wall thickness) can reveal the age of the tree and the prevailing climatic conditions during each season.

Glossary

apical meristem　顶端分生组织
cell division　细胞分裂
lateral meristem　侧生分生组织
herbaceous　草本的、叶状的
dicot　双子叶植物
monocot　单子叶植物
elongation　伸长、延长
axillary　腋生的、叶腋的
cambium　形成层
cork cambium　木栓形成层
tracheid　管胞
vessel　导管
sieve element　筛管
companion cell　伴胞
lignin　木质素
repel　击退、抵挡住、驱除
phelloderm　栓内层
periderm　周皮
lenticel　皮孔
annual ring　年轮

Lesson 9　Plant life cycles

The length of time from the beginning of development to the death of a plant is called the life cycle. The life cycle, on the other hand, is the sequence of stages a plant goes through from seed germination to seed production of the mature plant. Some plants, such as annuals, only need a few weeks to grow, produce seeds, and die. Other plants, such as the bristlecone pine, live for thousands of years. Some bristlecone pines have a documented age of 4,500 years. Even as some parts of a plant, such as regions containing meristematic tissue (the area of active plant growth consisting of undifferentiated cells capable of cell division) continue to grow, some parts undergo programmed cell death (apoptosis). The cork

found on stems and the water-conducting tissue of the xylem, for example, is composed of dead cells.

Annuals, Biennials, and Perennials

Plant species that complete their life cycle in one season are known as annuals, an example of which is *Arabidopsis*, or mouse-ear cress. Biennials, such as carrots, complete their life cycle in two seasons. In a biennial's first season, the plant has a vegetative phase, whereas in the next season, it completes its reproductive phase. Commercial growers harvest the carrot roots after the first year of growth and do not allow the plants to flower. Perennials, such as the magnolia, complete their life cycle in two years or more.

Monocarpic and Polycarpic Plants

In another classification based on flowering frequency, monocarpic plants flower only once in their lifetime; examples of monocarpic plants include bamboo and yucca. During the vegetative period of their life cycle (which may be as long as 120 years in some bamboo species), these plants may reproduce asexually, accumulating a great deal of food material that will be required during their once-in-a-lifetime flowering and setting of seed after fertilization. Soon after flowering, these plants die. Polycarpic plants form flowers many times during their lifetime. Fruit trees, such as apple and orange trees, are polycarpic; they flower every year. Other polycarpic species, such as perennials, flower several times during their life span, but not each year. By this method, the plant does not require all its nutrients to be channeled towards flowering each year.

Genetics and environmental conditions

As is the case with all living organisms, genetics and environmental conditions have a role to play in determining how long a plant will live. Susceptibility to disease, changing environmental conditions, drought, cold, and competition for nutrients are some of the factors that determine the survival of a plant. Plants continue to grow, despite the presence of dead

tissue, such as cork. Individual parts of plants, such as flowers and leaves, have different rates of survival. In many trees, the older leaves turn yellow and eventually fall from the tree. Leaf fall is triggered by factors such as a decrease in photosynthetic efficiency due to shading by upper leaves or oxidative damage incurred as a result of photosynthetic reactions. The components of the part to be shed are recycled by the plant for use in other processes, such as development of seed and storage. This process is known as nutrient recycling. However, the complex pathways of nutrient recycling within a plant are not well understood.

The aging of a plant and all the associated processes is known as senescence, which is marked by several complex biochemical changes. One of the characteristics of senescence is the breakdown of chloroplasts, which is characterized by the yellowing of leaves. The chloroplasts contain components of photosynthetic machinery, such as membranes and proteins. Chloroplasts also contain DNA. The proteins, lipids, and nucleic acids are broken down by specific enzymes into smaller molecules and salvaged by the plant to support the growth of other plant tissues. Hormones are known to play a role in senescence. Applications of cytokinins and ethylene delay or prevent senescence; in contrast, abscissic acid causes premature onset of senescence.

Glossary

life span 生命周期
annual 一年生植物
bristlecone pine 狐尾松
apoptosis 细胞凋亡
biennial 两年生植物
perennial 多年生植物
Arabidopsis 拟南芥
mouse-ear cress 鼠耳芥
carrot 胡萝卜
magnolia 木兰
monocarpic 单心皮的
bamboo 竹子
yucca 丝兰属植物
polycarpic 多心皮的
photosynthetic efficiency 光合效率
senescence 衰老
chloroplast 叶绿体
abscissic acid 脱落酸

Lesson 10 Natural and artificial methods of asexual reproduction in plants

Plants can undergo natural methods of asexual reproduction, performed by the plant itself, or artificial methods, aided by humans.

Natural methods of asexual reproduction

Natural methods of asexual reproduction include strategies that plants have developed to self-propagate. Many plants, such as ginger, onion, gladioli, and dahlia, continue to grow from buds that are present on the surface of the stem. In some plants, such as the sweet potato, adventitious roots or runners (stolons) can give rise to new plants. In *Bryophyllum* and *Kalanchoe*, the leaves have small buds on their margins. When these are detached from the plant, they grow into independent plants; they may also start growing into independent plants if the leaf touches the soil. Some plants can be propagated through cuttings alone.

Artificial methods of asexual reproduction

Artificial methods of asexual reproduction are frequently employed to give rise to new, and sometimes novel, plants. The methods include grafting, cutting, layering, and micropropagation.

Grafting

Grafting has long been used to produce novel varieties of roses, citrus species, and other plants. In grafting, two plant species are used: part of the stem of the desirable plant is grafted onto a rooted plant called the stock. The part that is grafted or attached is called the scion. Both are cut at an oblique angle (any angle other than a right angle), placed in close contact with each other, and are then held together. Matching up these two surfaces as closely as possible is extremely important because this will be holding the plant together. The vascular systems of the two plants grow and fuse, forming a graft. After a period of time, the scion starts producing shoots, eventually bearing flowers and fruits. Grafting is widely used in viticulture (grape growing) and the citrus industry. Scions capable of producing a particular fruit variety are grafted onto root stock with specific resistance to disease.

Cutting

Plants such as coleus and money plant are propagated through stem cuttings where a portion of the stem containing nodes and internodes is placed in moist soil and allowed to root. In some species, stems can start producing a root even when placed only in water. For example, leaves of the African violet will root if kept undisturbed in water for several weeks.

Layering

Layering is a method in which a stem attached to the plant is bent and covered with soil. Young stems that can be bent easily without any injury are the preferred plant for this method. Jasmine and bougainvillea (paper flower) can be propagated this way. In some plants, a modified form of layering known as air layering is employed. A portion of the bark or outermost covering of the stem is removed and covered with moss, which is then taped. Some gardeners also apply rooting hormone. After some time, roots will appear; this portion of the plant can be removed and transplanted into a separate pot.

Micropropagation

Micropropagation (also called plant tissue culture) is a method of propagating a large number of plants from a single plant in a short time under laboratory conditions. This method allows propagation of rare, endangered species that may be difficult to grow under natural conditions, are economically important, or are in demand as disease-free plants.

To start plant tissue culture, a part of the plant such as a stem, leaf, embryo, anther, or seed can be used. The plant material is thoroughly sterilized using a combination of chemical treatments standardized for that

species. Under sterile conditions, the plant material is placed on a plant tissue culture medium that contains all the minerals, vitamins, and hormones required by the plant. The plant part often gives rise to an undifferentiated mass, known as a callus, from which, after a period of time, individual plantlets begin to grow. They can be separated; they are first grown under greenhouse conditions before they are moved to field conditions.

Glossary

self-propagate 自行繁殖
ginger 姜
onion 洋葱
gladioli 唐菖蒲
dahlia 大丽花
adventitious 外来的
stolon 匍匐枝
Bryophyllum 落地生根属
Kalanchoe 伽蓝菜属
grafting 嫁接
micropropagation 微体繁殖
scion 幼枝
viticulture 葡萄栽培
node 节
internode 节间
African violet 非洲紫罗兰
jasmine 茉莉
bougainvillea 叶子花属、九重葛属
callus 愈伤组织
plantlet 幼小植物

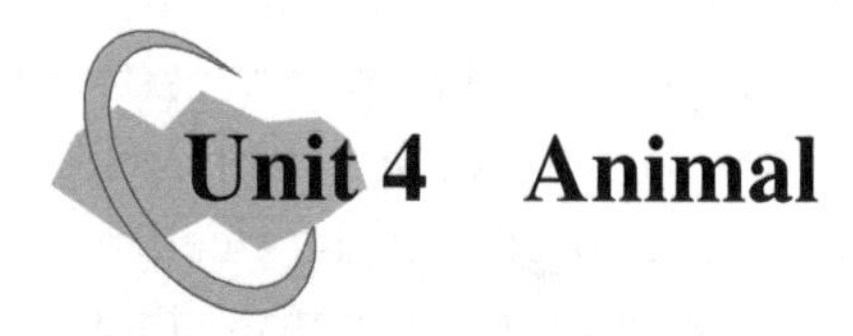

Unit 4 Animal

The word "animal" comes from the Latin animalis, referring only to non-human members of the kingdom Animalia. In 1874, Ernst Haeckel divided the animal kingdom into two subkingdoms according to cell number: Metazoa (multicellular animals) and Protozoa (single-celled animals). The protozoa are the simplest in the animal kingdom, but they are the most complex han any particular cell in higher organisms. They were later moved to the kingdom Protista, leaving only the metazoa. Thus Metazoa is now considered as a synonym of Animalia.

Animals are generally considered evolved from a flagellated eukaryote. Their closest known living relatives are the choanoflagellates, collared flagellates that have a morphology similar to the choanocytes of certain sponges. The first fossils are interpreted as being early sponges. They were found in 665 million-year-old rock. This unit mainly introduces animal characteristics, methods of animal reproduction and the constructing of an animal phylogenetic tree.

Lesson 11 Characteristics of the animal kingdom

Animal evolution began in the ocean over 600 million years ago with tiny creatures that probably do not resemble any living organism today. Since then, animals have evolved into a highly-diverse kingdom. Although over one million extant (currently living) species of animals have been identified, scientists are continually discovering more species as they explore ecosystems around the world. The number of extant species is estimated to be between 3 and 30 million.

But what is an animal? While we can easily identify dogs, birds, fish, spiders, and worms as animals, other organisms, such as corals and

sponges, are not as easy to classify. Animals vary in complexity, from sea sponges to crickets to chimpanzees, and scientists are faced with the difficult task of classifying them within a unified system. They must identify traits that are common to all animals as well as traits that can be used to distinguish among related groups of animals. The animal classification system characterizes animals based on their anatomy, morphology, evolutionary history, features of embryological development, and genetic makeup. This classification scheme is constantly developing as new information about species arises. Understanding and classifying the great variety of living species help us better understand how to conserve the diversity of life on earth.

Even though members of the animal kingdom are incredibly diverse, most animals share certain features that distinguish them from organisms in other kingdoms. All animals are eukaryotic, multicellular organisms, and almost all animals have a complex tissue structure with differentiated and specialized tissues. Most animals are motile, at least during certain life stages. All animals require a source of food and are, therefore, heterotrophic: ingesting other living or dead organisms. This feature distinguishes them from autotrophic organisms, such as most plants, which synthesize their own nutrients through photosynthesis. As heterotrophs, animals may be carnivores, herbivores, omnivores, or parasites. Most animals reproduce sexually with the offspring passing through a series of developmental stages that establish a fixed body plan. The body plan refers to the morphology of an animal, determined by developmental cues.

Animals with bilateral symmetry that live in water tend to have a fusiform shape: a tubular shaped body that is tapered at both ends. This shape decreases the drag on the body as it moves through water and allows the animal to swim at high speeds. Certain types of sharks can swim at fifty kilometers an hour, while some dolphins can swim at 32 to 40 kilometers per hour. Land animals frequently travel faster (although the tortoise and snail are significantly slower than sharks or dolphins). Another difference in the adaptations of aquatic and land-dwelling organisms is that aquatic organisms are constrained in shape by the forces of drag in the water since water has higher viscosity than air. However, land-dwelling organisms are constrained mainly by gravity; drag is relatively unimportant. For example, most adaptations in birds are for gravity, not for drag.

Most animals have an exoskeleton, including insects, spiders, scorpions, horseshoe crabs, centipedes, and crustaceans. Scientists estimate that, of insects alone, there are over 30 million species on our planet. The exoskeleton is a hard covering or shell that provides benefits to the animal, such as protection against damage from predators and from water loss (for land animals); it also provides for the attachments of muscles. As the tough and resistant outer cover of an arthropod, the exoskeleton may be constructed of a tough polymer, such as chitin, and is often biomineralized with materials, such as calcium carbonate. This is fused to the animal's epidermis. Ingrowths of the exoskeleton called apodemes function as attachment sites for muscles, similar to tendons in more advanced animals. In order to grow, the animal must first synthesize a new exoskeleton underneath the old one and then shed or molt the original covering. This limits the animal's ability to grow continually. It may limit the individual's ability to mature if molting does not occur at the proper time. The thickness of the exoskeleton must be increased significantly to accommodate any increase in weight. It is estimated that a doubling of body size increases body weight by a factor of eight. The increasing thickness of the chitin necessary to support this weight limits most animals with an exoskeleton to a relatively-small size.

The same principles apply to endoskeletons, but they are more efficient because muscles are attached on the outside, making it easier to compensate for increased mass. An animal with an endoskeleton has its size determined by the amount of skeletal system it needs in order to support the other tissues and the amount of muscle it needs for movement. As the body size increases, both bone and muscle mass increase. The speed achievable by the animal is a balance between its overall size and the bone and muscle that provide support and movement.

Glossary

creature 生物
extant species 现存物种
spider 蜘蛛
worm 蠕虫
coral 珊瑚
sponge 海绵动物
complexity 复杂
cricket 蟋蟀
chimpanzee 黑猩猩
anatomy 解剖
embryological 胚胎学的
genetic makeup 基因组成
heterotrophic 异养的
carnivore 食肉动物
herbivore 食草动物
omnivore 杂食动物
parasite 寄生生物
offspring 后代、子代
bilateral 左右对称的
fusiform 两端渐细的
dolphin 海豚
exoskeleton 外骨骼
scorpion 蝎子
horseshoe crab 马蹄蟹
crustacean 甲壳纲动物
arthropod 节肢动物

Lesson 12 Methods of animal reproduction

Reproduction (or procreation) is the biological process by which new "offspring" (individual organisms) are produced from their "parents". It is a fundamental feature of all known life that each individual organism exists as the result of reproduction. Most importantly, reproduction is necessary for the survival of a species. The known methods of reproduction are broadly grouped into two main types: sexual and asexual.

In asexual reproduction, an individual can reproduce without involvement with another individual of that species. The division of a bacterial cell into two daughter cells is an example of asexual reproduction. This type of reproduction produces genetically-identical organisms (clones), whereas in sexual reproduction, the genetic material of two individuals combines to produce offspring that are genetically different from their parents.

During sexual reproduction, the male gamete (sperm) may be placed inside the female's body for internal fertilization, or the sperm and eggs

may be released into the environment for external fertilization. Humans provide an example of the former, while seahorses provide an example of the latter. Following a mating dance, the female seahorse lays eggs in the male seahorse's abdominal brood pouch where they are fertilized. The eggs hatch and the offspring develop in the pouch for several weeks.

Asexual versus sexual reproduction

Organisms that reproduce through asexual reproduction tend to grow in number exponentially. However, because they rely on mutation for variations in their DNA, all members of the species have similar vulnerabilities. Organisms that reproduce sexually yield a smaller number of offspring, but the large amount of variation in their genes makes them less susceptible to disease.

Many organisms can reproduce sexually as well as asexually. Aphids, slime molds, sea anemones, and some species of starfish are examples of animal species with this ability. When environmental factors are favorable, asexual reproduction is employed to exploit suitable conditions for survival, such as an abundant food supply, adequate shelter, favorable climate, disease, optimum pH, or a proper mix of other lifestyle requirements. Populations of these organisms increase exponentially via asexual reproductive strategies to take full advantage of the rich supply resources. When food sources have been depleted, the climate becomes hostile, or individual survival is jeopardized by some other adverse change in living conditions, these organisms switch to sexual forms of reproduction.

Sexual reproduction ensures a mixing of the gene pool of the species. The variations found in offspring of sexual reproduction allow some individuals to be better suited for survival and provide a mechanism for selective adaptation to occur. In addition, sexual reproduction usually results in the formation of a life stage that is able to endure the conditions that threaten the offspring of an asexual parent. Thus, seeds, spores, eggs, pupae, cysts, or other "over-wintering" stages of sexual reproduction ensure the survival during unfavorable times as the organism can wait out adverse situations until a swing back to suitability occurs.

Glossary

reproduction 繁殖
parents 亲代
survival 幸存
sexual 有性的
fertilization 受精
seahorse 海马
abdominal 腹部的
pouch 育儿袋
exponentially 指数地
mutation 突变
vulnerability 缺陷、脆弱点
aphid 蚜虫类
slime mold 黏菌类
sea anemone 海葵
starfish 海星
jeopardize 危害
pupae 蛹

Lesson 13 Constructing an animal phylogenetic tree

Phylogenetic trees are constructed according to the evolutionary relationships that exist between organisms based on homologous traits.

Evolutionary trees, or phylogeny, are the formal study of organisms and their evolutionary history with respect to each other. Phylogenetic trees are most-commonly used to depict the relationships that exist between species. In particular, they clarify whether certain traits are homologous (found in the common ancestor as a result of divergent evolution) or homoplasy (sometimes referred to as analogous: a character that is not found in a common ancestor, but whose function developed independently in two or more organisms through convergent evolution). Evolutionary trees are diagrams that show various biological species and their evolutionary relationships. They consist of branches that flow from lower forms of life to the higher forms of life.

Evolutionary trees differ from taxonomy which is an ordered division of organisms into categories based on a set of characteristics used to assess similarities and differences. Evolutionary trees involve biological classification and use morphology to show relationships. Phylogeny is evolutionary history shown by the relationships found when comparing polymeric molecules such as RNA, DNA, or proteins of various organisms.

The evolutionary pathway is analyzed by the sequence similarity of these polymeric molecules. This is based on the assumption that the similarities of sequence result from having fewer evolutionary divergences than others. The evolutionary tree is constructed by aligning the sequences; the length of the branch is proportional to the amount of amino acid differences between the sequences.

Phylogenetic systematics informs the construction of phylogenetic trees based on shared characters. Comparing nucleic acids or other molecules to infer relationships is a valuable tool for tracing an organism's evolutionary history. The ability of molecular trees to encompass both short and long periods of time is hinged on the ability of genes to evolve at different rates, even in the same evolutionary lineage. For example, the DNA that codes for rRNA changes relatively slowly, so comparisons of DNA sequences in these genes are useful for investigating relationships between taxa that diverged a long time ago. Interestingly, 99% of the genes in humans and mice are detectably orthologous, and 50% of our genes are orthologous with those of yeast. The hemoglobin B genes in humans and in mice are orthologous. These genes serve similar functions, but their sequences have diverged since the time that humans and mice had a common ancestor.

Evolutionary pathways relating the members of a family of proteins may be deduced by examination of sequence similarity. This approach is based on the notion that sequences that are more similar to one another have had less evolutionary time to diverge than have sequences that are less similar. Evolutionary trees are used today for DNA hybridization, which determines the percentage difference of genetic material between two similar species. If there is a high resemblance of DNA between the two species, then the species are closely related. If only a small percentage is identical, then they are distantly related.

Animal phyla

The current understanding of evolutionary relationships between

animal, or metazoa, phyla begins with the distinction between "true" animals with true differentiated tissues, called eumetazoa, and animal phyla that do not have true differentiated tissues (such as the sponges), called parazoa. Both Parazoa and Eumetazoa evolve from a common ancestral organism that resembles the modern-day protists called choanoflagellates. These protist cells strongly resemble sponge choanocyte cells.

Eumetazoa are subdivided into radially-symmetrical animals and bilaterally-symmetrical animals and are classified into clade Radiata or bilateria, respectively. The cnidarians and ctenophores are animal phyla with true radial symmetry. All other Eumetazoa are members of the bilateria clade. The bilaterally-symmetrical animals are further divided into deuterostomes (including chordates and echinoderms) and two distinct clades of protostomes (including ecdysozoans and lophotrochozoans). Ecdysozoa includes nematodes and arthropods; named for a commonly-found characteristic among the group: exoskeletal molting (termed ecdysis). Lophotrochozoa is named for two structural features, each common to certain phyla within the clade. Some lophotrochozoan phyla are characterized by a larval stage called trochophore larvae, and other phyla are characterized by the presence of a feeding structure called a lophophore.

Glossary

phylogenetic tree 进化树
homologous 同源的
phylogeny 系统发育
convergent evolution 趋同进化
taxonomy 分类学
polymeric 聚合的，聚合体的
divergence 分歧
taxa 分类群
orthologous 直系同源
hemoglobin 血红蛋白、血红素
hybridization 杂交
metazoa 后生动物
phyla 类群
eumetazoa 真后生动物
parazoa 拟生动物
choanoflagellate 领鞭虫类
bilateria 两侧对称动物
cnidarian 刺胞动物
ctenophore 栉水母门动物
deuterostome 后口动物

chordate　脊索动物
echinoderm　棘皮动物
protostome　原口动物
nematode　线虫
arthropod　节肢动物
exoskeletal molting　体外骨骼蜕皮
ecdysis　换毛、蜕皮
lophotrochozoa　冠轮动物
trochophore larvae　担轮幼虫
lophophore　触手冠

Unit 5 Ecology

The word "ecology" ("Ökologie") was coined in 1866 by the German scientist Ernst Haeckel (1834–1919). Ecology is the scientific analysis and study of interactions among organisms and their environment. Within the discipline of ecology, researchers work at four specific levels, sometimes discretely and sometimes with overlap. These levels are organism, population, community, and ecosystem. In ecology, ecosystems are composed of dynamically-interacting parts, which include organisms, the communities they comprise, and the non-living (abiotic) components of their environment. This unit mainly introduces the basal contents of ecosystem.

Lesson 14 Food chains and food webs

In ecology, a food web describes the feeding connections between organisms in a biotic community. Both energy and nutrients flow through a food web, moving through organisms as they are consumed by an organism above them in the food web. A single path of energy through a food web is called a food chain.

Trophic levels

Each organism within a food web can be classified by trophic level according to their position within the web. Depending on an organism's location in a food web, it may be grouped into more than one of these categories. Energy and nutrients move up trophic levels in the following order: primary producers, primary consumers, secondary consumers, tertiary and other high-level consumers.

In both food webs and food chains, arrows point from an organism that is consumed to the organism that consumes it. In many ecosystems, the bottom of the food chain consists of photosynthetic organisms, such as plants or trophic level, known as primary producers. The organisms that consume the primary producers are herbivores: the primary consumers. Secondary consumers are usually carnivores that eat the primary consumers, while tertiary consumers are carnivores that eat other carnivores. Higher-level consumers feed on the next lower trophic levels, and so on, up to the organisms at the top of the food chain, which are called the apex consumers. Some lines within a food web may point to more than one organism; those organisms may occupy different trophic levels depending on their position in each food chain within the web.

The loss of energy in tropic levels

It is rare to find food chains that have more than four or five links because the loss of energy limits the length of food chains. At each trophic level, most of the energy is lost through biological processes such as respiration or finding food. Only the energy that is directly assimilated into an animal's consumable mass will be transferred to the next level when that animal is eaten. Therefore, after a limited number of trophic energy transfers, the amount of energy remaining in the food chain cannot support a higher trophic level. Although energy is lost, nutrients are recycled through waste or decomposition.

Types of food webs

Two general types of food webs are often shown interacting within a single ecosystem. As an example, a grazing food web has plants or other photosynthetic organisms at its base, followed by herbivores and various carnivores. A detrital food web consists of a base of organisms that feed on decaying organic matter (dead organisms), called decomposers or detritivores. These organisms are usually bacteria or fungi that recycle organic material back into the biotic part of the ecosystem as they themselves are consumed by other organisms. As all ecosystems require a

method to recycle material from dead organisms, most grazing food webs have an associated detrital food web. For example, in a meadow ecosystem, plants may support a grazing food web of different organisms, primary and other levels of consumers, while at the same time supporting a detrital food web of bacteria, fungi, and detrivorous invertebrates feeding off dead plants and animals.

food web 食物网
biotic community 生物群落
food chain 食物链
trophic level 营养级
primary producer 初级生产者
primary consumer 初级消费者
tertiary 第三的
herbivore 食草动物
carnivore 食肉动物
apex consumer 顶端消费者
assimilate 吸收
respiration 呼吸作用
decomposer 分解者
detritivore 食碎屑者
invertebrate 无脊椎动物

Lesson 15 Ecosystem dynamics

Ecosystem dynamics is the study of the changes in ecosystem structure caused by environmental disturbances or by internal forces. Various research methodologies measure ecosystem dynamics. Some ecologists study ecosystems using controlled experimental systems, while some study entire ecosystems in their natural state; others use both approaches.

Holistic ecosystem model

A holistic ecosystem model attempts to quantify the composition, interaction, and dynamics of entire ecosystems. A food web is an example of a holistic ecosystem model, which is the most representative of the ecosystem in its natural state. However, this type of study is limited by time and expense, as well as its limited feasibility to conduct experiments

on large natural ecosystems.

Experimental systems

For the reasons above, scientists study ecosystems under more controlled conditions. Experimental systems usually involve either partitioning a part of a natural ecosystem that can be used for experiments, termed a mesocosm, or by re-creating an ecosystem entirely in an indoor or outdoor laboratory environment, which is referred to as a microcosm. A major limitation to these approaches is that removing individual organisms from their natural ecosystem or altering a natural ecosystem through partitioning may change the dynamics of the ecosystem. These changes are often due to differences in species numbers and diversity, but also to environment alterations caused by partitioning (mesocosm) or re-creating (microcosm) the natural habitat. Thus, these types of experiments are not totally predictive of changes that would occur in the ecosystem from which they were gathered.

As both of these approaches have their limitations, some ecologists suggest that results from these experimental systems should be used only in conjunction with holistic ecosystem studies to obtain the most representative data about ecosystem structure, function, and dynamics.

Ecosystem models

Scientists use the data generated by these experimental studies to develop ecosystem models that demonstrate the structure and dynamics of ecosystems. Three basic types of ecosystem modeling are routinely used in research and ecosystem management: conceptual models, analytical models, and simulation models.

A conceptual model consists of flow charts to show interactions of different compartments of the living and nonliving components of the ecosystem. A conceptual model describes ecosystem structure and dynamics and shows how environmental disturbances affect the ecosystem, although its ability to predict the effects of these disturbances is limited.

Analytical and simulation models are mathematical methods of

describing ecosystems that are capable of predicting the effects of potential environmental changes without direct experimentation, although with limitations in accuracy. An analytical model is created using simple mathematical formulas to predict the effects of environmental disturbances on ecosystem structure and dynamics.

A simulation model is created using complex computer algorithms to holistically model ecosystems and to predict the effects of environmental disturbances on ecosystem structure and dynamics. Ideally, these models are accurate enough to determine which components of the ecosystem are particularly sensitive to disturbances. They can serve as a guide to ecosystem managers (such as conservation ecologists or fisheries biologists) in the practical maintenance of ecosystem health.

Glossary

ecosystem dynamics 生态系统动力学
methodology 方法论
approach 方法
holistic ecosystem mode 整体生态系统模型
mesocosm 中观
microcosm 微观
conceptual model 概念模型
analytical model 分析模型
simulation model 仿真模型
disturbance 干扰
mathematical methods 数学方法
algorithm 算法
ecosystem health 生态系统健康

Lesson 16 Climate change and biodiversity

Climate change, specifically, the anthropogenic (caused by humans) warming trend presently underway, is recognized as a major extinction threat, particularly when combined with other threats such as habitat loss. Scientists disagree about the probable magnitude of the effects, with extinction rate estimates ranging from 15 percent to 40 percent of species by 2050. Scientists do agree, however, that climate change will alter regional climates, including rainfall and snowfall patterns, making habitats less hospitable to the species living in them.

The warming trend will shift colder climates toward the north and south poles, forcing species to move with their adapted climate norms while facing habitat gaps along the way. The shifting ranges will impose new competitive regimes on species as they find themselves in contact with other species not present in their historic range. One such unexpected species contact is between polar bears and grizzly bears. Previously, these two species had separate ranges. Now, with their ranges are overlapping, there are documented cases of these two species mating and producing viable offspring. Changing climates also throw off species' delicate timing adaptations to seasonal food resources and breeding times. Many contemporary mismatches to shifts in resource availability and timing have recently been documented.

Range shifts are already being observed. For example, some European bird species' ranges have moved 91 km northward. The same study suggests that the optimal shift based on warming trends was double that distance, suggesting that the populations are not moving quickly enough. Range shifts have also been observed in plants, butterflies, other insects, freshwater fishes, reptiles, and mammals.

Climate gradients will also move up mountains, eventually crowding species higher in altitude and eliminating the habitat for those species adapted to the highest elevations. Some climates will completely disappear. The rate of warming appears to be accelerated in the arctic, which is recognized as a serious threat to polar bear populations that require sea ice to hunt seals during the winter months; seals are the only source of protein available to polar bears. A trend to decreasing sea ice coverage has occurred since observations began in the mid-twentieth century. The rate of decline observed in recent years is far greater than previously predicted by climate models.

Finally, global warming will raise ocean levels due to glacial melt and the greater volume of warmer water. Shorelines will be inundated, reducing island size, which will have an effect on many species; a number of islands will disappear entirely. Additionally, the gradual melting and subsequent refreezing of the poles, glaciers, and higher elevation mountains, a cycle that has provided freshwater to environments for centuries, will also be

jeopardized. This could result in an overabundance of salt water and a shortage of fresh water.

Glossary

anthropogenic 人为的
habitat 栖息地
rainfall 降雨量
snowfall 降雪量
grizzly bear 北美棕熊
offspring 后代、子代
butterfly 蝴蝶
reptile 爬行动物
crowding 拥挤现象
arctic 北极圈
population 种群
seal 海豹
global warming 全球变暖
shoreline 海岸线
melting 融化
glacier 冰川
freshwater 淡水
overabundance 过剩

Unit 6 Biotechnology

Biotechnology is the use of living systems and organisms to develop or make products, or "any technological application that uses biological systems, living organisms or derivatives thereof, to make or modify products or processes for specific use". For thousands of years, humankind has used biotechnology in agriculture, food production, and medicine. The term is largely believed to have been coined in 1919 by Hungarian engineer Károly Ereky. In the late 20th and early 21st century, biotechnology has expanded to include new and diverse sciences such as genomics, recombinant gene techniques, applied immunology, and development of pharmaceutical therapies and diagnostic tests. The unit mainly introduces basic techniques to manipulate genetic material, genetically modified organisms and basic techniques in protein analysis.

Lesson 17 Basic techniques to manipulate genetic material

Basic techniques used in genetic material manipulation include extraction, gel electrophoresis, PCR, and blotting methods.

To understand the basic techniques used to work with nucleic acids, remember that nucleic acids are macromolecules made of nucleotides (a sugar, a phosphate, and a nitrogenous base) linked by phosphodiester bonds. The phosphate groups on these molecules each have a net negative charge. An entire set of DNA molecules in the nucleus is called the genome. DNA has two complementary strands linked by hydrogen bonds between the paired bases. The two strands can be separated by exposure to high temperatures (DNA denaturation) and can be reannealed by cooling. The DNA can be replicated by the DNA polymerase enzyme. Unlike DNA,

which is located in the nucleus of eukaryotic cells, RNA molecules leave the nucleus. The most common type of RNA that is analyzed is the messenger RNA (mRNA) because it represents the protein-coding genes that are actively expressed.

DNA and RNA extraction

To study or manipulate nucleic acids, the DNA or RNA must first be isolated or extracted from the cells. This can be done through various techniques. Most nucleic acid extraction techniques involve steps to break open the cell and use enzymatic reactions to destroy all macromolecules that are not desired (such as degradation of unwanted molecules and separation from the DNA sample). Cells are broken using a lysis buffer (a solution that is mostly a detergent); lysis means “to split”. These enzymes break apart lipid molecules in the membranes of the cell and the nucleus. Macromolecules are inactivated using enzymes such as proteases that break down proteins, and ribonucleases (RNases) that break down RNA. The DNA is then precipitated using alcohol. Human genomic DNA is usually visible as a gelatinous, white mass. Samples can be stored at –80°C for years.

RNA analysis is performed to study gene expression patterns in cells. RNA is naturally very unstable because RNases are commonly present in nature and very difficult to inactivate. Similar to DNA, RNA extraction involves the use of various buffers and enzymes to inactivate macromolecules and preserve the RNA.

Gel electrophoresis

Because nucleic acids are negatively-charged ions at neutral or basic pH in an aqueous environment, they can be mobilized by an electric field. Gel electrophoresis is a technique used to separate molecules on the basis of size using this charge and may be separated as whole chromosomes or fragments. The nucleic acids are loaded into a slot near the negative electrode of a porous gel matrix and pulled toward the positive electrode at the opposite end of the gel. Smaller molecules move through the pores in the gel faster than larger molecules; this difference in the rate of migration separates the fragments on the basis of size. There are molecular-weight standard samples that can be run alongside the molecules to provide a size comparison. Nucleic acids in a gel matrix can be observed using various fluorescent or colored dyes. Distinct nucleic acid fragments appear as bands at specific distances from the top of the gel (the negative electrode end) on the basis of their size.

Amplification of nucleic acid fragments by polymerase chain reaction

Polymerase chain reaction (PCR) is a technique used to amplify specific regions of DNA for further analysis. PCR is used for many purposes in laboratories, such as the cloning of gene fragments to analyze genetic diseases, identification of contaminant foreign DNA in a sample, and the amplification of DNA for sequencing. More practical applications include the determination of paternity and detection of genetic diseases.

DNA fragments can also be amplified from an RNA template in a process called reverse transcriptase PCR (RT-PCR). The first step is to recreate the original DNA template strand (called cDNA) by applying DNA nucleotides to the mRNA. This process is called reverse transcription. This requires the presence of an enzyme called reverse transcriptase. After the cDNA is made, regular PCR can be used to amplify it.

Hybridization, southern blotting, and northern blotting

Nucleic acid samples, such as fragmented genomic DNA and RNA extracts, can be probed for the presence of certain sequences. Short DNA fragments called probes are designed and labeled with radioactive or fluorescent dyes to aid detection. Gel electrophoresis separates the nucleic acid fragments according to their size. The fragments in the gel are then transferred onto a nylon membrane in a procedure called blotting. The nucleic acid fragments that are bound to the surface of the membrane can then be probed with specific radioactively or fluorescently-labeled probe sequences. When DNA is transferred to a nylon membrane, the technique is called southern blotting; when RNA is transferred to a nylon membrane,

it is called northern blotting. Southern blots are used to detect the presence of certain DNA sequences in a given genome, and northern blots are used to detect gene expression.

Glossary

gel electrophoresis 凝胶电泳
blotting 印迹
nucleotide 核苷酸
phosphodiester 磷酸二酯
genome 基因组
denaturation 变性
polymerase enzyme 聚合酶
eukaryotic cell 真核细胞
macromolecule 大分子
lysis 融化 溶解
ribonuclease 核糖核酸酶
polymerase chain reaction 聚合酶链反应
contaminant 污染物
reverse transcriptase 逆转录酶
amplify 扩大，增加

Lesson 18 Genetically modified organisms

A genetically modified organism (GMO) is any organism whose genetic material has been altered using genetic engineering techniques, which can lead to the expression of desirable genes in plants and animals.

Herbert Boyer and Stanley Cohen made the first genetically modified organism (GMO) in 1973. They took a gene from a bacterium that provided resistance to the antibiotic kanamycin, inserted it into a plasmid and then induced another bacteria to uptake the plasmid. The bacteria were then able to survive in the presence of kanamycin. Boyer and Cohen expressed other genes in bacteria. This included genes from the toad Xenopus laevis in 1974, creating the first GMO expressing a gene from an organism from different kingdom.

GMOs are used in biological and medical research, production of pharmaceutical drugs, experimental medicine (e.g., gene therapy and vaccines against the Ebola virus), and agriculture (e.g., golden rice, resistance to herbicides), with developing uses in conservation. The term "genetically modified organism" does not always imply, but can include, targeted insertions of genes from one species into another. For example, a gene from a jellyfish, encoding a fluorescent protein called GFP, or green fluorescent protein, can be physically linked and thus co-expressed with mammalian genes to identify the location of the protein encoded by the GFP-tagged gene in the mammalian cell. Such methods are useful tools for biologists in many areas of research, including those who study the mechanisms of human and other diseases or fundamental biological processes in eukaryotic or prokaryotic cells.

Microbes

Bacteria were the first organisms to be modified in the laboratory, due to the relative ease of modifying their genetics. They continue to be important model organisms for experiments in genetic engineering. In the field of synthetic biology, they have been used to test various synthetic approaches, from synthesizing genomes to creating novel nucleotides. These organisms are now used for several purposes, and are particularly important in producing large amounts of pure human proteins for use in medicine.

Transgenic plants

Manipulating the DNA of plants (or creating genetically modified organisms called GMOs) has helped to create desirable traits, such as disease resistance, herbicide and pesticide resistance, better nutritional value, and better shelf-life. Plants are the most important source of food for the human population. Farmers developed ways to select for plant varieties with desirable traits long before modern-day biotechnology practices were established. Plants that have received recombinant DNA from other species are called transgenic plants. Because foreign genes can spread to other species in the environment, extensive testing is required to ensure ecological stability. Staples like corn, potatoes, and tomatoes were the first crop plants to be genetically engineered.

Transformation of plants using *Agrobacterium tumefaciens*

Gene transfer occurs naturally between species in microbial populations. Many viruses that cause human diseases, such as cancer, act by incorporating their DNA into the human genome. In plants, tumors caused by the bacterium *Agrobacterium tumefaciens* occur by transfer of DNA from the bacterium to the plant. Although the tumors do not kill the plants, they stunt the plants, which become more susceptible to harsh environmental conditions. Many plants, such as walnuts, grapes, nut trees, and beets, are affected by *Agrobacterium tumefaciens*. The artificial introduction of DNA into plant cells is more challenging than in animal cells because of the thick plant cell wall.

Researchers used the natural transfer of DNA from *Agrobacterium* to a plant host to introduce DNA fragments of their choice into plant hosts. In nature, the disease-causing *Agrobacterium tumefaciens* have a set of plasmids, called the Ti (Tumor inducing) plasmids (tumor-inducing plasmids), that contain genes for the production of tumors in plants. DNA from the Ti plasmid integrates into the infected plant cell's genome. Researchers manipulate the Ti plasmids to remove the tumor-causing genes and insert the desired DNA fragment for transfer into the plant genome. The Ti plasmids carry antibiotic resistance genes to aid selection and can be propagated in *Escherichia coli*. cells as well.

Transgenic animals

Although several recombinant proteins used in medicine are successfully produced in bacteria, some proteins require a eukaryotic animal host for proper processing. For this reason, the desired genes are cloned and expressed in animals, such as sheep, goats, chickens, and mice. Animals that have been modified to express recombinant DNA are called transgenic animals. Several human proteins are expressed in the milk of transgenic sheep and goats, while others are expressed in the eggs of chickens. Mice have been used extensively for expressing and studying the effects of recombinant genes and mutations.

Glossary

genetically modified organism 转基因生物
kanamycin 卡那霉素
toad Xenopus laevis 蟾蜍
pharmaceutical drug 医疗用药
Ebola virus 伊波拉病毒
herbicide 除草剂
jellyfish 水母
mammalian 哺乳动物
recombinant DNA 重组 DNA
Agrobacterium tumefaciens 根癌农杆菌
plasmid 质体

Lesson 19 Basic techniques in protein analysis

The ultimate goal of proteomics is to identify or compare the proteins expressed in a given genome under specific conditions, study the interactions between the proteins, and use the information to predict cell behavior or develop drug targets. Just as the genome is analyzed using the basic technique of DNA sequencing, proteomics requires techniques for protein analysis. The basic technique for protein analysis, analogous to DNA sequencing, is mass spectrometry.

Mass spectrometry

Mass spectrometry is used to identify and determine the characteristics of a molecule. It is a technique in which gas phase molecules are ionized and their mass-to-charge ratio is measured by observing acceleration differences of ions when an electric field is applied. Lighter ions will accelerate faster and be detected first. If the mass is measured with precision, then the composition of the molecule can be identified. In the case of proteins, the sequence can be identified. The challenge of techniques used for proteomic analyses is the difficulty in detecting small quantities of proteins, but advances in spectrometry have allowed researchers to analyze very small samples of protein. Variations in protein expression in diseased states, however, can be difficult to discern. Proteins are naturally-unstable molecules, which makes proteomic analysis much more difficult than genomic analysis.

X-ray crystallography and nuclear magnetic resonance

X-ray crystallography enables scientists to determine the three-dimensional structure of a protein crystal at atomic resolution. Crystallographers aim high-powered X-rays at a tiny crystal containing trillions of identical molecules. The crystal scatters the X-rays onto an electronic detector that is the same type used to capture images in a digital camera. After each blast of X-rays, lasting from a few seconds to several hours, the researchers precisely rotate the crystal by entering its desired orientation into the computer that controls the X-ray apparatus. This enables the scientists to capture in three dimensions how the crystal scatters, or diffracts, X-rays. The intensity of each diffracted ray is fed into a computer, which uses a mathematical equation to calculate the position of every atom in the crystallized molecule. The result is a three-dimensional digital image of the molecule.

Another protein imaging technique, nuclear magnetic resonance (NMR), uses the magnetic properties of atoms to determine the three-dimensional structure of proteins. NMR spectroscopy is unique in being able to reveal the atomic structure of macromolecules in solution, provided that highly-concentrated solution can be obtained. This technique depends on the fact that certain atomic nuclei are intrinsically magnetic. The chemical shift of nuclei depends on their local environment. The spins of neighboring nuclei interact with each other in ways that provide definitive structural information that can be used to determine complete three-dimensional structures of proteins.

Protein microarrays and two-hybrid screening

Protein microarrays have also been used to study interactions between proteins. These are large-scale adaptations of the basic two-hybrid screen. The premise behind the two-hybrid screen is that most eukaryotic transcription factors have modular activating and binding domains that can still activate transcription even when split into two separate fragments, as long as the fragments are brought within close proximity to each other. Generally, the transcription factor is split into a DNA-binding domain (BD)

and an activation domain (AD). One protein of interest is genetically fused to the BD and another protein is fused to the AD. If the two proteins of interest bind each other, then the BD and AD will also come together and activate a reporter gene that signals interaction of the two hybrid proteins.

Glossary

proteomics　蛋白组学
mass spectrometry　质谱分析
X-ray crystallography　X 射线晶体学
three-dimensional structure　三维结构
nuclear magnetic resonance　核磁共振
protein microarray　蛋白质微阵列
two-hybrid screen　双杂交筛选
DNA-binding domain　DNA 结合域
hybrid　杂种，混合物

Part II　Promotion Components

提升篇

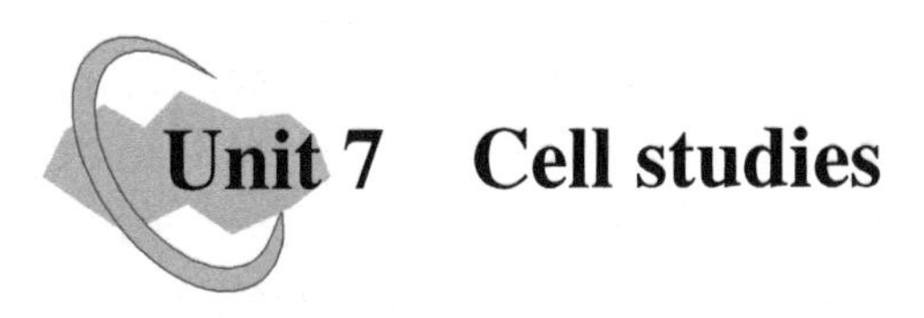

Unit 7 Cell studies

Lesson 20 Mechanisms of plant cell division

Cell division is a fundamentally important process, allocating cellular contents to two daughter cells. In eukaryotes, this task is in large accomplished by the highly dynamic cytoskeleton. While many cellular structures and molecular processes are conserved between eukaryotes, plant cells exhibit unique features that might be attributed to their mostly sedentary life style. Moreover, most plant cells are immobilized within the network of rigid polymeric cell walls. The microtubule cytoskeleton is the main driver of cell division forming the chromosome segregating spindle, which is the common mitotic feature shared by eukaryotes. However, plant cells accomplish this task by a self-organizing acentrosomal mechanism. In addition, two specialized cytoskeletal structures are instrumental in the selection and maintenance of the division plane and in the formation of the cell plate. The preprophase band (PPB) is a remarkable belt-like cortical formation of microtubules and actin filaments assembling at the G2 to M transition, outlining the periphery of the future division plane. However, not all cell types (i.e., endosperm, meiocytes) utilize the PPB to determine the division site, proposing that alternative strategies exist. The other plant-specific cytoskeletal array is the phragmoplas, which physically partitions daughter cell contents by synthesis of the cell plate. Motor and non-motor microtubule-associated proteins (MAPs) reorganize microtubules by means of sliding, cross-linking or severing into the characteristic functional arrays in accordance to cell cycle progression. Thus vigorous coordination of cellular signaling, cytoskeletal organization and cell cycle regulators are required for cell division.

Recently, a number of reports have started to fill some of the gaps that we face in understanding plant cell division. Regulation of division plane selection is still poorly understood on a molecular level and it is yet too early to deduce whether asymmetric divisions are regulated by common mechanisms that operate in different tissue context or whether tissue-specific polarity cues trigger asymmetry. Nevertheless, the available data suggest that once the division plane is defined, the universal molecular machinery is employed to progress through mitosis. Furthermore, the use of evolutionary distant plant models and even different cell types demonstrates that divers strategies evolved in plant cells to achieve the accurate insertion of the new cell wall. For instance, it is yet to be determined whether actin-myosin-and microtubule-kinesin-based phragmoplast guidance mechanisms act entirely independently or whether they converge on the same pathway at some point. Further exciting challenges include how the acentrosomal microtubule cytoskeleton is organized into the plant-specific mitotic arrays that might also bear relevant mechanistic insight for other biological kingdoms. The latest reports also show that plant cell division research is increasingly benefiting from recent advances in light microscopy and more is to be expected by application of the rapidly evolving plane illumination technology to continuously improve 3D and 4D resolution. Furthermore, computation allows taking a fresh look at longstanding hypothesis and puts them to the test. (Lipka E, Herrmann A, Mueller S, 2015)

Lesson 21 Browning phenomena in plant cell cultures

Browning phenomena are ubiquitous in plant cell cultures and have been found in many species, such as *Thalictrum minus*, grape, Scots pine, *Hevea brasiliensis*, tobacco, Sorghum, *Aralia cordata*, *Glycyrrhiza inflata* and *Taxus*. The browning phenomena severely hamper a widespread application of plant cell cultures and scientific studies with plant cells as typical materials. Plant cell culture technology is a promising alternative for production of high-value secondary metabolites, and this technology

has been extended to produce heterologous proteins. But the application of this biotechnology has been hampered by browning problems that occur during plant cell cultures, including callus induction, cell suspension and scale-up culture.

The ubiquitous browning severely impedes the plant cell cultures by affecting cell growth and causing cell death. However, up to now, the browning mechanisms in plant cell culture are still unclear and there have been no ideal control methods. To solve this unfathomed problem, researchers usually passively add enzyme inhibitors, antioxidants and adsorbents in medium, or shorten subculture time to decrease the browning degree. But all these measures cannot fundamentally prevent browning. The additives are usually toxic to plant cells, and have to be removed from cultures in the later period.

Just because of the unclear browning mechanisms in plant cell cultures, this problem still has not been well controlled. However, there are several studies on the browning of fruits and vegetables which may give us some clues for further studies in plant cell cultures. In foods, browning phenomena can be divided into enzymatic browning and non-enzymatic browning, and the enzymatic browning is attributed to the oxidation of phenolic compounds by polyphenol oxidase (PPO). The major groups of reactions leading to browning are enzymatic browning, and the non-enzymatic browning is favored by heat treatments. Enzymatic browning phenomena in foods are usually induced by wounding resulting from mechanical stress. PPOs are localized in chloroplasts or other non-green plastids, while phenolics are located in the vacuole. Browning reactions are only initiated when plants suffer wounding, where the loss of cellular compartmentation allows PPO and phenolics to meet. Phenolic compounds, as the substrates of PPOs, are derivatives of the phenylpropanoid pathways in plants. Phenolic compounds, ubiquitous in plants, show a large diversity of structures, including phenolic acids (vanillin, gallic acid, caffeic acid, ferulic acid, p-coumaric acid and etcetera), and polyphenols such as flavonoids and polymers derived from these various groups. In addition, lignin of polyphenolic macromolecules and its precursors of monolignols all belong to phenolics. Owing to the large diversity of phenolics in plants, the specific phenolic compounds that mainly contribute to the browning in plants were also unrevealed. (Dong YS, Fu CH, Su P, et al. 2016)

Lesson 22 Cell mechanics

21st century biomechanics research has entered an exciting era of investigation; where the mechanical behaviors of cells and tissues can be both a direct consequence, and a regulating factor of biological function and cellular architecture. The underlying goal of current cell biomechanics research is to combine theoretical, experimental, and computational approaches to construct a realistic description of cell mechanical behaviors that can be used to provide new perspectives on the role of mechanics in disease.

Cell mechanics research has great potential to provide new perspectives on pathologies and classic biological research questions. To facilitate wider use of mechanical experimental tools and cell rheological characterization we have outlined a simple set of considerations for non-experts in the field looking to attempt mechanical measurements. The choice of experimental tool depends on the lengthscale of the sample and the level of force that is needed to deform the sample. Many techniques are now spatially accurate to the subcellular level and sensitive enough to measure pico to nano newton levels of force and can be scaled appropriately by altering the size and stiffness of the measurement probe. There are also a variety of environmental and experimental conditions that need to be considered, such as temperature control and the interface with other measurement techniques, such as optical microscopy. Upon choosing the mechanical measurement tool that will be spatially accurate, can apply the correct forces and comply with physiological condition, there are also a variety of mechanical loading protocols that can be employed. Considering the cell active features, the loading condition and the timescale of mechanical measurement could have a direct relevance in probing the underlying active cellular processes. Typical loading protocols involve step changes in stress or strain while monitoring the ensuing relaxation response. Other loading protocols include application of oscillatory

mechanical excitations that provide significant insight about time dependent mechanical properties.

While their ability to describe the complex cell rheological behavior is extremely limited, linear viscoelastic characterization of cellular mechanical responses in terms of spring-dashpot models lead to estimation of stiffnesses and viscosities that are useful for evaluating cell mechanics under different biological and chemical perturbations. Beyond simple linear mechanical descriptions are scale free models that better explain some of the commonly observed universal cell mechanical behaviors. These can be applied with great effect to capture some of the mechanical responses of the cell under different loading conditions and at a wide range of timescales. Other reconstituted gel and biphasic models provide a mechanistic insight about cell rheological measurements. However, a unifying theory that describes all of the complexities of cell mechanical behavior remains an exciting and active area of research. (Moeendarbary E Harris AR, 2014)

Lesson 23 Stem cell research: trends and perspectives on the evolving international landscape

Stem cell research is an exciting yet complex and controversial science. The field holds the potential to revolutionize the way human diseases are treated, and many nations have therefore invested heavily in stem cell research and its applications. However, human stem cell research is also controversial with many ethical and regulatory questions that impact a nation's policies.

Stem cells in China

We also examined the publication trends of China and the United States specifically, to see whether we can observe the impact of country level policy decisions in the publication data.

China is a country which shows steady growth in stem cell research supported by its major funding initiatives. In 2001, the Chinese Ministry of Science and Technology (MOST) launched two independent stem cell programs followed by a number of funding initiatives intended to further promote stem cell research, applications and public awareness. At the same time, China has been working to strengthen ethical guidelines and regulations. In total, the national government's stem cell funding commitment is estimated at more than 3 billion RMB (close to 500 million USD) for the 5 year period from 2012 to 2016. Confronted with the healthcare needs of a rapidly aging population of nearly 1.4 billion, the impetus behind much stem cell research, so far, has understandably been clinical translation and development.

Looking at the publication data from our study, we see that stem cell publications have grown from representing 0.2% of China's total publication output in 2001 to a peak of 0.82% in 2008, followed by a marginal decline to 0.76% by 2012. As also observed globally, China's iPS cell publication output surpassed hES cell publication output in 2010, after which hES cell output shows fluctuation.

Science policy and Human Embryonic Stem Cell (hES) research in the USA

The United States is an interesting case study, as reported in our study, they are the world leader in stem cell research considering that they produce the highest absolute publication volume, as well as high relative activity levels, indicating a high focus on stem cell research, and show high field-weighted citation impact. Yet, they have had to grapple with the practical and ethical dilemmas that are inherent in this field, and changing views of different administrations, as governments changed.

The result is a series of policy changes, some of which limited federal funding for hES cell research, while others loosened the limitations. Servellen and Oba map such policies along with the corresponding publication output (relative to total country output). Despite the restrictive policies between 2001 and 2009, the United States show steady output growth, which has been supported through individual, state and industry

funding as well as donations. We do observe changes in hES cell publication output that coincide with changes in regulation. While such changes in publication output are probably not best explained using a one factor model, these findings are hardly surprising, as we expect science policy to greatly impact scientific activity. Such an analysis can provide insight into the degree to which science policy has indeed affected publication output.

In recent years, stem cell research has grown remarkably, showing a growth rate more than double the rate of world research publications from 2008 to 2012. However, this increase is not uniform across all stem cell research areas. Our analysis showed that both ES and hES fields have grown more slowly than the stem cell field overall. In contrast, iPS cell publications have shown explosive growth, as would be expected of a new and promising field of research, and iPS cell publication volumes surpassed that of hES cell publications in 2010. However, both cell types continue to be highly active areas.

Stem cell research has attracted considerable attention within the scientific community: stem cell publications overall are cited 50% more than all other publications in related disciplines, while ES cell publications are cited twice the world rate, and iPS cell publications nearly three times the world rate. This high-growth, high-impact field encompasses research across many cell types, with a focus ranging from the most fundamental to the clinical. Reflecting the field's ongoing development and clinical promise, approximately half of all stem cell publications are associated with regenerative medicine or drug development, a trend that is particularly pronounced in iPS cell research.

Stem cell research is developing fast, with some experimental pluripotent stem cell treatments already in clinical trials. Active debates are underway to adapt regulatory frameworks to address the specific challenges of developing, standardizing, and distributing cell-based therapies, while advances in basic research continue to provide a fuller understanding of how stem cells can be safely and effectively used. Cell replacement or transplantation therapies are not the only application of stem cell research: already the first steps are being taken towards use of cells derived from pluripotent stem cells, in drug discovery and testing. It is with great interest and anticipation that we watch the further development of this exciting field of science. (Servellen SV, Oba I, 2014)

Lesson 24 Suicide of aging cells prolongs life span in mice

They are lurking in your heart, your liver, your kidneys, and maybe even your brain: run-down cells that could be making you age. A new study of mice shows that spurring these so-called senescent cells to self-destruct extends the animals' lives and delays some aspects of aging.

"It's a landmark paper," says cell and molecular biologist Francis Rodier of the University of Montreal in Canada, who wasn't connected to the study. "It's providing biological evidence that senescence is involved in the aging process."

Cells senesce after suffering DNA damage or other types of stress. Although they remain alive, senescent cells lose the ability to divide. Researchers think that this cellular birth control evolved to thwart the formation of tumors, but it provides other benefits as well. The stagnant cells release chemicals that help wounds heal, for instance. But senescence also causes harm. If stem cells stop dividing, organs can deteriorate because they can't replace cells that have died. Furthermore, the chemicals released by senescent cells can damage surrounding tissues and, to researchers' surprise, promote tumor growth.

Our bodies build up senescent cells as we get older, and researchers have been trying to nail down their impact on aging for more than 50 years. Cancer biologist Jan van Deursen of the Mayo Clinic College of Medicine in Rochester, Minnesota, and colleagues took a direct approach to the problem, genetically engineering a strain of fast-aging mice so that their senescent cells committed suicide in response to a drug. In 2011, the researchers revealed that pruning senescent cells from the mice slowed their physical breakdown as they got older—although it didn't extend their

lives.

To find out if the approach worked in rodents other than the fast-aging mice, van Deursen and colleagues genetically modified two other mouse strains to kill their own senescent cells after receiving the same drug. When the rodents reached middle age, the researchers began injecting them with the compound twice a week. Although the procedure couldn't eliminate senescent cells from the animals, it could kill 50% to 70% of them in some tissues, van Deursen says.

After undergoing the treatment for 6 months, the mice were healthier in many ways than a set of control animals. As the scientists report online today in *Nature*, pruning senescent cells reduced the amount of damage to the blood-filtering structures in the kidneys. The animals' hearts were better able to cope with stress than were the hearts of control mice. Even the behaviors of the treated mice were different. They were more daring and youthful than the control mice. Like middle-aged folks who'd rather watch TV than hit the clubs, the controls were less active and more reluctant to explore new environments.

But the finding that grabbed the researchers' attention was that destroying senescent cells boosted the average life span of the two mouse strains by more than 20%. Some of the increased longevity may have stemmed from a beneficial effect on cancer. Removal of senescent cells didn't prevent tumors from forming in the rodents, but it did slow their growth. "It had more of an impact on life span than I would have predicted," van Deursen says.

Not all age-related problems in the mice improved, however. Their memory, muscle strength, coordination, and balance—all of which decline as we grow older—were no better than those of control rodents. Deleting senescent cells doesn't spare the animals from aging entirely, van Deursen says. "It has an attenuating effect. You still get age-dependent changes, and the mice still die."

"This study is a big step toward validating the approach of targeting senescent cells," says cell biologist Christian Sell of Drexel University College of Medicine in Philadelphia, Pennsylvania, who wasn't connected to the research.

Scientists have identified other approaches that slow aging in experimental animals, such as deleting certain genes or drastically cutting calories, notes geneticist Ned Sharpless of the University of North Carolina School of Medicine in Chapel Hill. However, says Sharpless, who wasn't connected to the research but did help found a company that makes a diagnostic test for senescent cells, these approaches are impractical for humans. For example, some would require a person to take a drug for decades to see only a small effect, he says. But deleting senescent cells could be feasible in people, he says. For the first time, a researcher can say, "if I can figure out a way to kill senescent cells with a small molecule or an antibody, I could do a clinical trial."

In fact, clinical trials might not be that far off. The mice in the study were genetically altered to respond to the drug, but a company that van Deursen co-founded and a separate group of researchers have already discovered compounds that can kill senescent cells in unmodified mice. It might soon be possible to test whether removing these cells can forestall age-related diseases, such as atherosclerosis, that cause so much suffering as we get older, van Deursen says. "We accumulate senescent cells, and they take away healthy years." (Leslie M, 2016)

Lesson 25 Nobel honors discoveries on how cells eat themselves

Nobel Prize in Physiology or Medicine in 2006 has been awarded to Yoshinori Ohsumi, a cell biologist at the Tokyo Institute of Technology's Frontier Research Center, for his work on autophagy, the process in which cells degrade and recycle cellular components.

"Ohsumi's discoveries led to a new paradigm in our understanding of how the cell recycles its content," a press release by the Nobel Assembly at the Karolinska Institute (KI) in Stockholm says. "His discoveries opened the path to understanding the fundamental importance of autophagy in

many physiological processes, such as in the adaptation to starvation or response to infection. Mutations in autophagy genes can cause disease, and the autophagic process is involved in several conditions including cancer and neurological disease."

"Of course for a researcher, there is no higher honor," Ohsumi said this evening at a hastily called press conference on the campus of the Tokyo Institute of Technology.

"This is an excellent decision," biochemist Volker Haucke of the Leibniz Institute for Molecular Pharmacology said at a meeting this morning in Berlin, where scientists watched the announcement live. "With Ohsumi, they have awarded the prize to a scientist who investigated a phenomenon in yeast that was seen as a side phenomenon, but that turned out to be central to molecular medicine," Haucke said. "It's very well deserved. ...He is a prime example of someone who did basic research and discovered a process that otherwise might have stayed hidden for decades."

In the 1950s and 1960s, researchers recognized that some animal cells use autophagy to recycle proteins and other cellular machinery. They knew that the process was especially active when the cell was under stress, for example when nutrients were in short supply or when the organism was fighting off an infection. But how the process worked—and even which cells used the method—was unclear.

Ohsumi and his colleagues set out to explore whether yeast, a single-celled organism that nevertheless uses many of the same biochemical processes as animal cells, could help answer some of the outstanding questions. ("I thought of trying something others weren't working on, so I started research into yeast," he explained at his press conference.) Ohsumi developed strains of yeast that lacked key enzymes suspected of playing a role in autophagy, hoping to see what happened to the cells when the process didn't work as it should. When they starved the yeast, the scientists found that the cells developed unusually large vacuoles, the cellular garbage dumps that collect materials to be recycled. Usually yeast vacuoles were too small to see under the light microscope, but in the mutant yeast, they grew so large they were easy to observe.

Ohsumi then used chemicals to induce more mutations in the yeast strains, looking for cells that failed to form visible vacuoles even when they were starving. Such cells, he reasoned, lacked genes that were important for autophagy to work properly. In a key paper published in FEBS Letters in 1993, Ohsumi and and his team identified 15 essential genes involved in the process. Further studies showed that very similar genes controlled the process in animal and human cells, and also helped piece together how the genes work together to keep the cell's recycling centers running.

Since then, Ohsumi and others have shown that autophagy plays a crucial role in embryo development, cell differentiation, and the immune system. A breakdown in autophagy can lead to a wide variety of diseases, including cancer, diabetes, and Huntington disease. A healthy autophagy system is correlated with longevity, and a faulty one can accelerate symptoms of aging.

"One reason autophagy has become so important is the realization that it is a highly regulated process, not just some automatic breakdown," says Peter-Michael Kloetzel, head of the Laboratory for Proteolytic Systems at the Charité university medical center in Berlin. As Daniel Klionsky, a researcher at the University of Michigan, Ann Arbor, explains it in this playful art and science video about autophagy, "the cell carries out a dance of spring cleaning 365 days a year."

"I think Ohsumi is the right person" to win the Nobel, says David Rubinsztein, who studies the role of autophagy in neurodegenerative diseases at the University of Cambridge Institute for Medical Research in the United Kingdom. "While there are many other people who have made important contributions to the field, he is justifiably considered the father of the field,"he says. "His lab was the first to identify yeast genes that regulate autophagy. Those discoveries have allowed us to then understand how autophagy is important in mammalian systems, because the yeast genes are very well conserved." (Science News Staff, 2016)

Unit 8 Microorganism studies

Lesson 26 Human adaptation to arsenic-rich environments

Many organisms have adapted to tolerate toxic chemicals in their environments; however, we know little about human adaptation to toxic chemicals. Exposure to inorganic arsenic is associated with multiple severe health effects, including increased morbidity and mortality in early life, cancer, cardiovascular and liver toxicity, and probably diabetes. In a few regions in the world, such as the Andean highlands, human populations have lived for thousands of years with drinking water contaminated with arsenic. This raises the question as to whether such populations may have adapted over time to their toxic environment.

The efficiency of arsenic metabolism strongly affects susceptibility to arsenic toxicity. In the body, cellular enzymes methylate inorganic arsenic to monomethylarsonic acid (MMA) and then dimethylarsinic acid (DMA). The fraction of arsenic present as MMA shows a positive association with arsenic toxicity, indicating that MMA is more toxic than DMA. By contrast, DMA is more readily excreted in urine and expelled from the body. The fractions of arsenic metabolites in human urine vary in different populations (fraction of MMA: 2%～30%). Indigenous populations in the Andes, including in the Argentinean village of San Antonio de los Cobres (SAC), show uniquely low urinary excretion of MMA.

The enzyme arsenic (+3 oxidation state) methyltransferase (AS3MT) plays a key role in arsenic methylation. Polymorphisms in AS3MT are associated with the arsenic methylation as shown in several populations, e.g., in Bangladesh, Argentina and Mexico. In particular, the AS3MT alleles associated with efficient arsenic methylation vary markedly in frequency. Individuals from SAC and surrounding villages have higher frequencies of inferred protective AS3MT haplotypes that other Native American and Asian populations. This observation led to the hypothesis that natural selection has favored AS3MT haplotypes that associate with more efficient arsenic metabolism in populations that have lived with arsenic exposure for many generations. Schlebusch et al. performed a genome-wide association study (GWAS) using dense, genome-wide markers and well-characterized arsenic metabolism phenotypes to demonstrate that AS3MT is likely to be the leading gene for arsenic methylation in humans. They found that in the people from SAC, the genomic region around AS3MT shows dramatic signs of selection, indicating adaptation to arsenic-rich environments. (Schlebusch CM, Gattepaille LM, Engström K, et al. 2015)

Lesson 27 Your poor diet might hurt your grandchildren's guts

Here's another reason to eat your vegetables. Trillions of microbes in the human large intestine—known as the microbiome—depend on dietary fiber to thrive and give us energy. As fiber intake declines, so, too, does the range of bacteria that can survive in the gut. Now, a new study of multiple generations of mice fed a low-fiber diet indicates that this diversity plummets further with each generation, a hint of what might be happening in the human gut as we continue eating a contemporary diet of refined foods. The work might also help explain rises in many Western diseases, such as inflammatory bowel disease and obesity.

"This is a seminal study," says microbial ecologist Jens Walter, of the University of Alberta in Canada. "The magnitude by which the low-fiber diet depletes the microbiome in the mouse experiments is startling."

For much of human history in hunter-gatherer and early agrarian times, daily fiber intake was likely at least three or four times the officially recommended amounts today and several times greater than average U.S.

consumption now (about 15 grams). The trend has led many researchers, including microbiologist Erica Sonnenburg of Stanford University in Palo Alto, California, to suspect that the well-documented low diversity of gut microbes among people in developed countries—some 30% less diverse than in modern hunter-gatherers—is, in part, a product of drastically reduced fiber intake.

The new study confirms this relationship in rodents. Sonnenburg and her colleagues raised mice in a germ-free environment and then fed them human feces, giving them human gut bacteria. When fed a low-fiber diet (about 30% less fiber than the control high-fiber chow), the animals experienced a substantial dip in gut microbial diversity (with about 60% of the microbes losing at least half their populations). Mice then kept on low-fiber food and allowed to breed produced offspring with an even lower diversity. And subsequent generations of low-fiber-fed mice continued to lose whole groups of microbes as the bacteria reached such reduced numbers in the parents that they could not be passed on via birth, nursing, or even mice's tendency to eat each other's feces.

By the fourth generation, the mouse microbiotas seemed to have reached a new normal, a stable low-diversity microbiota, harboring only just more than a quarter of the diversity enjoyed by the first generation. Notably, none of the low-fiber generations could be "rescued" by reintroducing high-fiber food. To reach the bacterial variety of their great-grandparents, the mice needed a fecal transplant from a high-fiber group and a high-fiber diet.

The findings raise questions of how our own diet affects our offspring. "While we pass on relatively few changes in our human DNA for each generation, this study indicates that we are potentially passing on huge changes in our gut microbiome," Sonnenburg says. Because the mice in this study started with microbes from a Westerner, whose microbiome was already low in diversity, the study also suggests that, "it may be possible for the Western microbiota to lose additional diversity," she adds. Microbiologist Eric Martens of the University of Michigan Medical School in Ann Arbor, marvels at the extent of the loss. "The surprise is that a proportion of organisms can actually be driven to extinction," he says.

Researchers have yet to prove that the same rapid decrease in microbe diversity is occurring over human generations—and if it is, just what it means for health. "In these complex ecosystems it is very difficult to know the exact outcome of biodiversity loss," Sonnenburg says. But, she notes, "it is likely these extinctions within the microbiota would have big effects." For example, other research has found that obese people are more likely to have lower microbial diversity in their guts than lean peers. And their studies of more macroenvironments teach us that a diverse ecosystem is heartier and quicker to rebound than a less diverse one. (Katherine Harmon Courage, 2016)

Lesson 28 Antibiotic use and its consequences for the normal microbiome

For 70 years, antibiotics have been a pillar of medicine and are being used worldwide on an enormous scale. In many countries, antibiotic use exceeds one course per capita per year. In 2010, the top seven antibiotic classes were consumed in an estimated 70 billion individual doses, which equates to about 10 pills, capsules, or teaspoons for every man, woman, and child on earth, an annual rate that appears to be rising. This magnitude of use is based at least in part on the perception, among both health professionals and the public, that antibiotics are completely safe. We all are aware of mild, self-limited problems, such as rashes and drug reactions, and doctors know about serious but very rare side effects, yet at a functional level, most of us consider these risks so close to zero that they do not usually factor into the equation about use. There also is the cost of antibiotic resistance, but because it predominantly affects the community rather than the treated individual, its avoidance does not usually affect clinical judgments about whether or not treatment should be given. Parents would rather have their ill child treated with a drug they see as safe and effective than worry about the impact of that single course on the future of humankind, and their doctors generally agree.

However, this construct of essentially complete personal safety is

illusory. Shortly after antibiotics began to be used to treat ill people and animals, farmers discovered that adding low doses of antibiotics to the food or water of their livestock would promote their growth; the earlier in life the exposure began, the more profound the effect. This observation alone provides an important clue that antibiotic exposure affects metabolic development. Antibiotic use has been widespread because it leads to growth promotion and, therefore, increased profit for farmers. But does this massive decades-long worldwide "experiment" on the farm teach us anything about human health? Is it possible that the antibiotics that we give our children early in life to treat their infections—whether severe (uncommon) or mild (very common)—are influencing a critical window in the development of their own metabolism?

In recent years, scientists have been exploring this question, with mostly consistent results. Observational, clinical, and epidemiologic studies focused on young children are providing a growing body of evidence that antibiotic exposure is associated with increased risk for a variety of diseases including obesity, types 1 and 2 diabetes, inflammatory bowel diseases, celiac disease, allergies, and asthma. Experimental models are providing increasing evidence that these associations are not just correlative but are causal. Studies in mice have found that antibiotic exposure, by disrupting the development of the early-life microbiome, which often causes loss of species and strain diversity, leads to metabolic perturbations that affect adiposity and bone growth and alter normal immunologic development.

A variety of evidence indicates that the risks appear greatest for young children. Paradoxically, the perinatal period through the first 2 years of life is the time when per capita antibiotic usage is most intensive. An emerging concept to consider is that the effects of antibiotics may be cumulative in an individual, with both epidemiologic and experimental data supporting that view. One hypothesis is that antibiotic courses may lead to species loss, especially for taxa that were low in number at that time, yet which may have important metabolic functions. The problem would be most important for taxa with unique functions, although this may not happen in all patients or with all courses. Because we each inherit much of our microbiome from our mother, a further hypothesis is that environmental impacts on the microbiome (including antibiotic and dietary exposures) are cumulative across generations. We need to carefully assess these hypotheses, because the implications are substantial.

Yet antibiotics are vital for health care. It is difficult to imagine optimal health without an umbrella of antibiotics to use when needed. Nevertheless, practitioners have not been taking the biologic cost of antibiotic use into account sufficiently in making treatment decisions. Differences in perceptions about how risk-free antibiotic treatment is may in part account for the enormous variation in rates of their use from practitioner to practitioner, between localities and across countries. The emergence of awareness of the biologic costs of a treatment surely requires modulation of its usage. We must clearly understand the real costs, including the differences between particular antimicrobial agents in their effects on the microbiome and, thus, the consequent sequelae on child development.

Rather than carpet-bombing germs into submission with broad-spectrum antibiotics, we will need more laserlike approaches to develop drugs against specific pathogens, minimize damage to essential symbiotic microbial species, and preserve community structure and function in the healthy (and developing) microbiome. Future research, based on our extensive and growing knowledge of bacterial genes and genomes, should aim to develop truly narrow-spectrum agents, each ideally targeting a single pathogen. This strategy also requires new diagnostic tests that accurately and economically differentiate between bacterial and viral infections, among specific bacteria, and also distinguish between colonization and infection. Host-specific, indeed individual, differences may require more personalized approaches to antibacterial therapies. Shortening treatment courses is another approach that needs more research, as do the complex trade-offs that arise between emergence of resistance and collateral damage.

There always will be instances in which children must be treated with an antibiotic, but the collateral effects could be mitigated. Should we bank every healthy child's fecal specimen, so we can chase each antibiotic

course with a dose of their own pretreatment microbiota? Or by studying microbiota before and after antibiotic treatment, can we identify a few key organisms to replace in that child and then prescribe well-characterized pharmaceutical-grade standard strains, the probiotics of the future, to be administered in the appropriate vehicle?

We also are learning that other drugs are affecting the microbiome in ways that were not anticipated. For example, metformin, widely used in the treatment of type 2 diabetes, has important effects on microbial populations. It may be that metformin's antimicrobial actions, which in turn affect short-chain fatty acid metabolism, determine the efficacy of the drug, rather than its direct effects on tissues. However, no drugs are as important as antibiotics because of their usage in virtually all children worldwide and because their activities, specifically targeted to bacteria, strongly select and shape microbial community structure. Even later in life, antibiotic exposures may have consequences in terms of risks for metabolic and neoplastic diseases and for acquisition of resistant organisms.

The third and fourth edges of the antibiotic sword—cost to the community and to a person's future health—are both being driven by antibiotic overuse. First, we must control those excesses, but mitigation will only stabilize the situation not reverse the deterioration that likely has progressively occurred with socioeconomic development. Ultimately, we may need to recover the biodiversity lost as a result of these generations of antibiotic use (and other insults). The goal should be to restore the status quo ante, using probiotics, perhaps with accompanying prebiotics, to replace vital missing and/or extinct species and strains that modulate crucial developmental pathways. This critically important next scientific frontier in human health will require much research. (Blaser MJ, 2016)

Lesson 29 The global ocean microbiome

Oceanographers began studying the ocean microbiome in earnest over four decades ago, when it was recognized that microbes are responsible for nearly all of the energy flux in this largest and most dilute biological system on Earth. Much has been learned about the microbes that play key roles in every marine element cycle, but much is still unknown about the factors regulating their activity. Although the number of marine microbes per liter of seawater reaches into the billions, their small size means that, statistically, each microbe is separated by 100 to 200 body lengths from its closest neighbors. Yet recognition of microscale structuring of both microbial communities and marine organic matter suggests that the ocean microbiome does not operate as stand-alone cells in a watery soup.

Several decades of 16*S* ribosomal RNA gene analysis has revealed distinct and recurring bacterial communities in the ocean. More recent characterizations of marine archaea, protists, and viruses are filling out the taxonomic inventories of the ocean microbiome and showing that membership is predictable over seasons, ocean depth, and organic matter features. The retrieval of proteorhodopsin—a gene that allows cells to harvest energy from sunlight without complex photosynthetic machinery—from an uncultured ocean microbe marked the first exciting discovery from the use of "meta-omics" methodologies in the ocean. Now these techniques are the central tools for converting inventories of organisms and functions into explicit linkages between the two. Substantial progress has been made toward unraveling how and where microbes participate in ocean biogeochemical processes, as well as toward recognizing new categories of nonpredatory microbial alliances that operate based on the exchange of nitrogen, vitamins, hormones, and antibiotics.

Several characteristics of the ocean microbiome distinguish it from microbiomes on or in animals, plants, and soils. First, the primary producers that fuel the ocean are exclusively microbial and thus are a part of the microbiome. This is the case for photosynthesis in the surface ocean and for chemosynthesis carried out in deeper waters. The ocean microbiome is responsible for half of all primary production occurring on Earth. Second, trophic categories are particularly difficult to assign in the ocean microbiome, with no clear division of organisms into canonical autotrophic and heterotrophic roles. Proteorhodopsin, anoxygenic phototrophy, and chemolithotrophic energy acquisition from inorganic compounds create trophic mayhem among members of the ocean microbiome. Having multiple strategies for meeting metabolic requirements may be an advantage in this chemically dilute and physically dynamic environment.

Last, heterogeneity in the structure of seawater organic matter has become a foundational concept for the ocean microbiome because it aligns with differences in microbial attributes. Bacteria and archaea that live singly in seawater differ from those that intermingle on the various marine polymer networks and organic surfaces in terms of phylogenetic affinity, metabolism, and capabilities for motility, chemotaxis, and defense. Single bacteria and archaea are numerically dominant in terms of cells, genes, and transcripts, but those clustered near surfaces have higher per-cell rates of metabolism and growth. The importance of material exchanges and signaling networks between neighboring cells in the ocean, as well as the consequences spatial arrangements impose on biogeochemical processing, are not yet understood.

Earth's changing climate is predicted to decrease carbon fixation by microbial primary producers, favor smaller picophytoplankton over larger nano- and microphytoplankton, and impose stress on photosynthetic microbes that form calcium carbonate shells. The structure of phytoplankton communities, in turn, has implications for the abundance and composition of organic substrates for heterotrophic microbes, as well as for dictating which trophic strategies will be under selection in the future ocean. Taking stock of the ocean microbiome in terms of cells, genes, transcripts, and proteins now has a long tradition in oceanography. Linking these stocks with the regulation of critical ecosystem functions is the next challenge. One key step in this process is the identification of the molecules that pass between microbes as substrates, nutrients, signaling molecules, and defensive compounds; these are the "currencies" of ocean microbiome function. (Moran MA, 2015)

Lesson 30 Gut microbes give anticancer treatments a boost

Checkpoint inhibitors, which aim to unleash the power of the immune system on tumors, are some of the most impressive new cancer treatments. But most patients who receive them don't benefit. Two new studies of mice suggest a surprising reason why these people may not have the right mixture of bacteria in their guts. Both studies demonstrate that the composition of the gut microbiome—the swarms of microorganisms naturally dwelling in the intestines—determines how effective these cancer immunotherapies are.

The studies are the first to link our intestinal denizens to the potency of checkpoint inhibitors, drugs that thwart one of cancer's survival tricks. To curb attacks on our own tissues, immune cells carry receptors that dial down their activity. But tumor cells can also stimulate these receptors, preventing the immune system from attacking them. Checkpoint inhibitors like ipilimumab—which has been on the market since 2011—nivolumab, and pembrolizumab stop tumor cells from stimulating the receptors.

The new work could change how doctors use the drugs. "Both of the papers show convincingly that microbes can affect the treatments," says immunologist Yasmine Belkaid of the National Institute of Allergy and Infectious Disease in Bethesda, Maryland, who wasn't connected to the new studies. In the past, researchers have typically looked for mutations in patients' genomes that might explain why a particular checkpoint inhibitor isn't working, says molecular biologist Scott Bultman of the University of North Carolina School of Medicine in Chapel Hill. The new results are encouraging, he says, because "it's easier to change your gut microbiota than your genome."

Checkpoint inhibitors can shrink tumors and extend patients' lives, sometimes by years. Yet only a fraction of recipients improve. About 20% of melanoma patients treated with ipilimumab live longer, for example. Researchers don't know what distinguishes them from the other 80%.

A side effect of the drug steered oncoimmunologist Laurence Zitvogel of the Gustave Roussy Cancer Campus in Villejuif, France, and colleagues toward the microbiome. Ipilimumab often triggers colitis, an inflammation of the large intestine, where part of our microbiome lives. That side effect suggests checkpoint inhibitors and the microbiome interact. Following up on that possibility, the researchers tracked the growth of tumors implanted in mice lacking intestinal bacteria. The checkpoint inhibitor they tested was less powerful in the animals.

Further analysis by Zitvogel and colleagues suggested that certain bacteria in the *Bacteroides* and *Burkholderia* genera were responsible for the antitumor effect of the microbiome. To confirm that possibility, the researchers transferred the microbes into mice that had no intestinal bacteria, either by feeding the microorganisms to the animals or giving them the *Bacteroides*-rich feces of some ipilimumab-treated patients. In both cases, an influx of microbes strengthened the animals' response to one checkpoint inhibitor. "Our immune system can be mobilized by the trillions of bacteria we have in our gut," Zitvogel says.

Immunologist Thomas Gajewski of the University of Chicago (UC) in Illinois and colleagues came to a similar conclusion after noticing a disparity between mice they had obtained from two suppliers. Melanoma tumors grew slower in mice from Jackson Laboratory than in mice from Taconic Farms. The microbiomes of rodent cagemates tend to homogenize—the animals eat each other's feces—so the researchers housed mice from both suppliers together. Cohabitation erased the difference in tumor growth, indicating it depends on the types of microbes in the rodents' guts.

When they analyzed the microbiomes of the mice, the researchers pinpointed a bacterial genus known as the *Bifidobacterium*. The team found that feeding mice from Taconic Farms a probiotic that contains several Bifidobacterium species increased the efficiency of a checkpoint inhibitor against tumors. "The endogenous antitumor response is significantly influenced by your commensal bacteria," says co-author Ayelet Sivan, who was a Ph.D. student at UC when the research was conducted. Both groups reported their results online today in Science.

The two teams implicated different bacterial groups, but that doesn't worry microimmunologist Christian Jobin of the University of Florida College of Medicine in Gainesville. "Different drugs, different bugs, but the same endpoint," he says. He adds that the new work complements a pair of 2013 studies that demonstrated that the microbiome affects how well chemotherapy works.

The discovery "opens up novel ways to potentially augment therapy," says Cynthia Sears, an infectious disease specialist at Johns Hopkins

School of Medicine in Baltimore, Maryland. For example, it might be possible to beef up a patient's antitumor response with probiotics. But researchers also see some potential roadblocks. As Zitvogel notes, regulatory agencies in the United States and Europe haven't approved the use of probiotics for cancer patients. Also unclear is how the microbes boost the immune response—gut bacteria are key to the immune system's development, but researchers aren't sure how they tweak its function in mature animals. And scientists are just learning how to tinker with the microbiome. "It is not clear that we can meaningfully manipulate the microbiota and create positive health effects," Sears says. Nonetheless, researchers say, the studies suggest that we may have some powerful new allies in the fight against cancer. (Leslie M, 2015)

Lesson 31 Zika virus kills developing brain cells

As fear of the Zika virus spreads nearly as quickly as the pathogen itself, two new laboratory studies offer the first solid evidence for how it could cause brain defects in babies: The virus appears to preferentially kill developing brain cells. The observation bolsters the growing case for a connection between the virus, which is spreading rapidly across Latin America, and an increase in the number of cases of microcephaly, a birth defect in which the brain fails to grow properly. The new work, done independently by two groups, shows that the virus readily infects neural stem cells—the precursors of neurons and other brain cells—whether they are grown on cell culture plates or coaxed to form 3D minibrains called cerebral organoids.

The work "is going to be very important," says Madeline Lancaster, a developmental biologist who studies human brain development at the Medical Research Council Laboratory of Molecular Biology in Cambridge, U.K. The results "are quite consistent with what you're seeing in the babies with microcephaly."

Zika virus, named after a forest in Uganda where it was first isolated decades ago, usually causes only mild symptoms in people, including fever and rash. But after the virus started spreading across northeastern Brazil

last year, doctors there noticed a striking increase in the number of babies born with microcephaly. Many of the mothers reported having symptoms consistent with Zika infection during their pregnancies. But it has been difficult to prove a link between the virus and the birth defects because blood tests for Zika virus are only accurate for about a week after infection.

Nevertheless, circumstantial evidence has accumulated. Researchers have identified the virus in amniotic fluid of pregnant women whose fetuses were diagnosed with microcephaly and also in the brain tissue of a fetus diagnosed with the disorder. But because researchers had conducted scant research on the virus before this year, they had little data to suggest how the virus could cause such damage.

To gauge the virus's possible effects on the developing brain, researchers at Johns Hopkins University in Baltimore, Maryland, and Florida State University in Tallahassee used induced pluripotent stem (iPS) cells to grow, in lab dishes, immature brain cells called human cortical neural progenitor cells. (iPS cells are adult cells that have been reprogrammed into stem cells that can grow into most of the tissues in the body.) They then exposed the neural progenitor cells to a lab strain of Zika virus.

The virus readily infected the neural stem cells, neuroscientists Hongjun Song and Guo-li Ming, virologist Hengli Tang, and their colleagues report today in *Cell Stem Cell.* Three days after the virus was applied, 85% of the cells in the culture dishes were infected. In contrast, when the virus was applied to cultures of fetal kidney cells, embryonic stem cells, and undifferentiated iPS cells, it infected fewer than 10% of the cells by day 3. Immature neurons derived from the neural progenitor cells were also less susceptible to the virus; 3 days after receiving a dose of the virus, fewer than 20% of those cells were infected.

The researchers noticed that the infected progenitor cells were not killed right away. Instead, the virus "hijacked the cells," using the cellular machinery to replicate themselves, Song says. That helped the virus to spread quickly through the cell population, he says. His team also reports that infected cells grew more slowly and had interrupted cell division cycles, which could also contribute to microcephaly.

In a separate set of experiments, other researchers found that the virus can hamper the growth of another type of neural stem cell. In a preprint posted online on 2 March, neuroscientist Patricia Garcez and stem cell researcher Stevens Rehen at the D'Or Institute for Research and Education in Rio de Janiero, Brazil, report growing human iPS cells into clusters of neural stem cells called neurospheres, as well as into 3D organoids that in some ways resemble a miniature version of the human brain. When they infected the growing cells with Zika virus isolated from a Brazilian patient, the virus quickly killed most of the neurospheres and left the few survivors small and misshapen. Infected organoids grew to less than half their normal size.

Lancaster says the results echo earlier studies of gene mutations that cause microcephaly, which also affect neural progenitor cells. "You have two very different causes of microcephaly, but you see something very similar happening: a depletion of neural stem cells, and that would lead to fewer neurons" in the developing brain, she says.

Plenty of questions about the Zika virus and its apparent link to birth defects remain unanswered. Both Garcez and Song say they are now repeating their experiments with other viruses, including dengue, a virus closely related to Zika that is prevalent in the regions currently affected by the outbreak. (Some scientists suspect that previous exposure to other viruses could affect the outcome of Zika infections.) Researchers also still need to figure out how the virus crosses the placenta and infects the fetus directly, something most viruses can't do. (Vogel G, 2016)

Unit 9 Plant studies

Lesson 32 Abiotic and biotic stress combinations

Owing to their sessile lifestyle, plants are continuously exposed to a broad range of environmental stresses. The main abiotic stresses that affect plants and crops in the field are being extensively studied. They include drought, salinity, heat, cold, chilling, freezing, nutrient, high light intensity, ozone (O_3) and anaerobic stresses. Nevertheless, field conditions are unlike the controlled conditions used in the laboratory. Under natural conditions, combinations of two or more stresses, such as drought and salinity, salinity and heat, and combinations of drought with extreme temperature or high light intensity are common to many agricultural areas around the world and could impact crop productivity. A comparison of all major U.S. weather disasters that exceeded a billion dollars each, between 1980 and 2012, indicates that a combination of drought and heat stress caused extensive agricultural losses of $200 billion. By contrast, over the same period, drought alone caused $50 billion worth of damage to agricultural production. In addition, current climate prediction models indicate a gradual increase in ambient temperature, and an enhancement in the frequency and amplitude of heat stress in the near future. Moreover, high temperatures will be accompanied by other weather disasters, such as extended droughts, that could drastically impact crop production worldwide. An urgent need to generate crops with enhanced tolerance to stress combinations therefore exists.

In addition to abiotic stresses, under natural conditions, plants face the threat of infection by pathogens (including bacteria, fungi, viruses and nematodes) and attack by herbivore pests. The habitat range of pests and pathogens can be influenced by climate changes. For example, increasing temperatures are known to facilitate pathogen spread. Moreover, many abiotic stress conditions were shown to weaken the defense mechanisms of plants and enhanced their susceptibility to pathogen infection. Major crops growing in our future fields are therefore likely to be exposed to a greater range and number of abiotic and biotic conditions, as well as their combination.

Because different stresses are most likely to occur simultaneously under field conditions, a greater attempt must be made to mimic these conditions in laboratory studies. In 2002 and 2004, it was revealed that the molecular response of plants to a combination of drought and heat stress is unique and cannot be directly extrapolated from the response of plants to drought or heat stress applied individually. Since then, many studies have uncovered the responses of plants to different combinations of stresses involving drought, salt, extreme temperature, heavy metals, UV-B, high light, O_3, CO_2, soil compaction and biotic stresses. These studies demonstrated that, despite a certain degree of overlap, each stress condition required a unique mechanism of response, tailored to the specific needs of the plant, and that each combination of two or more different stresses may also require a specific response. In addition, the simultaneous occurrence of different biotic and abiotic stresses was shown to result in a high degree of complexity in plant responses, as the responses to these combined stresses are largely controlled by different signaling pathways that may interact and inhibit one another. Metabolic and signaling pathways involved in the response of plants to stress combination were found to include transcription factors, photosynthesis, antioxidant mechanisms, pathogen responses, hormone signaling, and osmolyte synthesis. However, the majority of the mechanisms underlying the tolerance of plants to stress combinations are still unknown and further studies are required to address them. (Suzuki N, Rivero RM, Shulaev V, et al. 2014)

Lesson 33 Does climate directly influence NPP globally ?

Understanding climate effects is critical to anticipate the impacts of climate change on ecosystem processes, such as carbon storage, which have feedbacks on the climate system. According to current textbooks on physiology and ecology, climate has strong direct impacts on net primary productivity (NPP) in ecosystems worldwide. Many studies have established empirical relationships between NPP and climate within and across ecosystems. There are well-established trends for the increase of NPP across ecosystems with higher temperature and precipitation from tundra to the temperate zone and sometimes from the temperate zone to the tropics, and evidence for the influence of climate change on specific ecosystems is accumulating weekly. Most approaches to examining relationships between NPP and climate across ecosystems considered the total effect of climate variables on NPP, that is, both direct effects and those that are indirect, via differences in stand structure. Thus, the direct effects of climate include its influence on the physiological and environmental determinants of plant growth, which are directly regulated by temperature and water availability, including the kinetics of photosynthetic rates and respiration rates and plant biomass allocation. Indirect effects of climate on NPP include the influence of climate on the structure of given ecosystems. For example, NPP is strongly influenced by total stand biomass (M_{tot}), and stands are often larger where rainfall is higher.

Resolving and disentangling the direct and indirect effects of climate on NPP is clearly important, because direct effects would signify more immediate responses to ongoing climate change. However, a recent paper has challenged the consensus, arguing, based on a statistical analysis of a global data set for woody plant-dominated communities, that climate has negligible direct influences on NPP at the global scale. They explained this conclusion as the result of plants' convergence in productivity despite global climate variation due to selection to maximize plant growth across climate gradients. They hypothesized that climate may have indirect effects, although these were not presented or quantified.

The need for rigorous analyses of climate impacts has never been more crucial. Current textbooks state that climate directly influences ecosystem annual net primary productivity (NPP), emphasizing the urgent need to monitor the impacts of climate change. A recent paper challenged this consensus, arguing, based on an analysis of NPP for 1247 woody plant communities across global climate gradients, that temperature and precipitation have negligible direct effects on NPP and only perhaps have indirect effects by constraining total stand biomass (M_{tot}) and stand age (a). The authors of that study concluded that the length of the growing season (l_{gs}) might have a minor influence on NPP, an effect they considered not to be directly related to climate. In the article published by Chu et al., they describe flaws that affected that study's conclusions and present novel analyses to disentangle the effects of stand variables and climate in determining NPP. They re-analyzed the same database to partition the direct and indirect effects of climate on NPP, using three approaches: maximum-likelihood model selection, independent-effects analysis, and structural equation modeling. These new analyses showed that about half of the global variation in NPP could be explained by M tot combined with climate variables and supported strong and direct influences of climate independently of M tot , both for NPP and for net biomass change averaged across the known lifetime of the stands (ABC = average biomass change). They show that l_{gs} is an important climate variable, intrinsically correlated with, and contributing to mean annual temperature and precipitation, all important climatic drivers of NPP. Their analyses provide guidance for statistical and mechanistic analyses of climate drivers of ecosystem processes for predictive modeling and provide novel evidence supporting the strong, direct role of climate in determining vegetation productivity at the global scale. (Chu C, Bartlett M, Wang Y, et al. 2015)

Lesson 34 The effects of enhanced UV-B radiation on plants

In spite of the current efforts to restrict the production of ozone depleting substances, thinning of the stratospheric ozone layer and increased penetration of ultraviolet-B (UV-B) radiation to the earth's surface will continue for decades. Enhanced UV-B is one of the most important abiotic stress factors that have many direct and indirect effects on plants, and enhanced UV-B was generally characterized as being harmful to plants.

Many researches suggested that the harmful effects of UV-B radiation on plants were often a consequence of reactive oxygen species (ROS) production. Consequently, some experiments were conducted to try to alleviate the damage of enhanced UV-B on plants by some measures. Mahdavian et al. reported that foliar spray of salicylic acid counteracted the UV-B effects on pepper pigment contents. Nitric oxide enhanced plant UV-B protection by scavenging ROS and up-regulating gene expression of the phenylpropanoid biosynthetic pathway. Research indicated that appropriate amounts of selenium could reduce ROS content and membrane lipid peroxidation, and increase the antioxidant ability in wheat, ryegrass, lettuce, strawberries and buckwheat.

In recent years, however, some interesting results about UV-B beneficial effects on secondary metabolism processes in medicinal plants have been found. Most of the medically active ingredients in medicinal plants are secondary metabolites. However, active ingredients content in many medicinal plants is not high enough to have a health-promoting effect. At present, a few studies have reported UV-B radiation effects on medicinal plants. The experiment performed by Kumari et al. showed that enhanced UV-B radiation significantly increased the essential oil compositions and total phenolics content in sweet flag (*Acoruscalamus*). Kumari and Agrawal studied the changes in leaf morphology, physiology and secondary metabolites of lemon grass (*Cymbopogoncitratus*) induced

by enhanced UV-B radiation. Manukyan found that low dose of UV-B radiation increased polyphenol content of lemon catmint (*Nepetacataria*), lemon balm (*Melissa officinalis* L.) and sage (*Salvia officinalis* L.). These studies demonstrate that enhanced UV-B radiation could induce secondary metabolism processes, and increase active ingredients content in medicinal plants. (Yao XQ, Chu JZ, Ma CH, et al. 2015; Yao XQ, Chu JZ, He XL, et al. 2014)

Lesson 35 Sulfur deficiency–induced repressor proteins optimize glucosinolate biosynthesis in plants

Sulfur, as an essential macronutrient, plays a crucial role in plant growth and development. Photosynthetic organisms use sulfate (SO_4^{2-}) as a primary sulfur source to synthesize an array of S-containing metabolites, including the amino acids cysteine (Cys) and methionine (Met), the tripeptide glutathione (GSH), vitamins and cofactors (such as thiamine, biotin, and coenzyme A), and chloroplastic sulfolipids. Moreover, primary sulfur assimilation is a prerequisite for synthesizing glucosinolates (GSLs) in *Brassicales*.

GSLs are nitrogen and sulfur-containing compounds found in the *Brassicaceae* family, including several important crops, such as oilseed rape (*Brassica napus*/*Brassica rapa*), cabbage (*Brassica oleracea* var. *capitata*), broccoli (*B. oleracea* var. *italica*), Chinese cabbage (*B. rapa*), and the model plant *Arabidopsis thaliana*. GSLs are important defense compounds against pathogens and herbivores and also act as S-storage sources. Moreover, potential health benefits of GSL-rich diets to humans come from the carcinopreventive properties of GSL hydrolysis products, which have been documented in multiple studies. For example, sulforaphane, an isothiocyanate derivative of 4-methylsulfinylbutyl GSL, and other isothiocyanates are potential candidates to prevent tumor growth by blocking the cell cycle, and have a potential for treating *Helicobacter pylori*-caused gastritis and stomach cancer.

GSLs are divided into three groups depending on their amino acid

precursors: aliphatic, benzenic (or aromatic), and indolic GSLs. In *A. thaliana*, 40 structurally different GSLs have been detected, most of which are aliphatic and indolic GSLs derived from Met and tryptophan (Trp), respectively. Because of their importance in agriculture and for human health, the GSL biosynthetic pathway was extensively investigated, and more than 20 genes involved in GSL biosynthesis have been identified in *Arabidopsis* to date. Despite our knowledge on the GSL biosynthetic pathways, understanding of the regulatory mechanisms and their synthesis upon environmental perturbations, for example, S deficiency (−S), remains fragmentary. Several R2R3 MYB family transcription factors have been identified as positive regulators of GSL synthesis, that is, MYB28, MYB29, and MYB76 as those of aliphatic GSLs and MYB34, MYB51, and MYB122 as those of indolic GSLs. Among them, MYB28 and MYB34 were identified as the dominant regulators of aliphatic and indolic GSLs, respectively, and are considered the major transcriptional inducers of GSL biosynthetic genes. In addition, basic helix-loop-helix (bHLH) transcription factors MYC2 (bHLH06), MYC3 (bHLH05), and MYC4 (bHLH04) have been reported to regulate GSL levels, to some extent, both dependent and independent of the aforementioned MYB transcription factors.

GSL accumulation is responding to plant development and abiotic factors, such as nitrogen and sulfur supply. In plants of the Brassicaceae family, the backbone of GSLs contains three S atoms, which can account for up to 30% of the total sulfur content of the entire plant. Thus, GSLs need to be tightly regulated in relation to the S nutritional status. Under short-term sulfur starvation (−S), all the major GSL biosynthetic genes, such as *MAM*, *CYP79*, and *CYP83* families, are down-regulated, and consequently, the GSL levels decrease. Concomitantly with the down-regulation of GSL synthesis, up-regulation of GSL catabolic genes under −S is reported by several transcriptomic studies. The S-storage function of GSL has been shown by the disruption of GSL transporters in *Arabidopsis* seeds, eliminating seed-borne GSLs and resulting in reduced seedling growth under −S. Thus, it can be hypothesized that, under −S, GSLs provide an important sulfur source and that the plant prioritizes protein synthesis and other essential functions above defense. However, it is largely unknown how sulfate availability regulates the expression of GSL pathway genes.

In addition to the modification of GSL metabolism, plants increase sulfate uptake and sulfur assimilation capacity in response to −S. The EIL family transcription factor SLIM1 has been identified as a regulator of plant −S responses associated with the up-regulation of sulfate uptake, GSL catabolism, and the down-regulation of GSL synthesis. This broad range of the metabolic pathways regulated by SLIM1 suggests the existence of other protein factors that mediate −S signals specific to each metabolic pathway. The presence of additional mediators is also suggested based on the fact that the expression levels of the MYB transcription factors do not fully correlate with the decreased GSL levels under –S. *MYB28* is not down-regulated under –S, and its expression level becomes even higher under long-term sulfur starvation, whereas *MYB29* and *MYB76* are repressed in both early and late phases of –S. This provides an implication that an additional mechanism may be involved in the negative regulation of GSL biosynthesis under –S.

Several transcriptome studies have revealed that a set of functionally unknown genes termed S-marker genes, including *sulfur deficiency induced 1* (*SDI1*; At5g48850) and *SDI2* (At1g04770), are up-regulated under sulfur starvation. A homologous gene of *SDI* has also been identified by a complementary DNA-based amplified fragment length polymorphism (cDNA-AFLP) analysis of field-grown, S-deficient wheat (*Triticum aestivum*). *SDI1* may play a functional role in the utilization of stored sulfate under −S because *Arabidopsis sdi* knockout lines accumulate more sulfate than do wild-type (WT) controls. SLIM1 appears to down-regulate *SDI1* and *SDI2* under S-sufficient conditions and the opposite is required for the induction of *SDI1* and *SDI2* under –S, whereas its functionality is probably modulated through a posttranslational mechanism, which can be more complex and possibly involve additional factors *SDI1* and *SDI2* levels are correlated to the *O*-acetylserine (OAS) content under a wide range of stress conditions, among them is sulfate starvation. OAS is the precursor of Cys synthesis in the S assimilation pathway and accumulates

under –S conditions. Hence, it has been considered to be a signaling compound for −S responses. (Aarabi F, kusajima M, Tohge T, et al. 2016)

Lesson 36 Fine roots—functional definition expanded to crop species

A recent review by McCormack *et al.* proposes to split fine roots (roots < 2 mm diameter) into two different functional groups: Absorptive and Transport. This is a significant step forward to account for some of the previous comments by Pregitzer and Zobel on the situation with fine roots. The historic designation of fine roots as roots < 2 mm diameter was arbitrary (i.e. different species might, rationally, have had different cut-off points) and originated with research on long-lived perennials, that is, trees. In many crop species (both annual and perennial), however, 2-mm-diameter roots are often considered large to very large.

The McCormack *et al.* review primarily addresses relatively long-lived perennials to the exclusion of crop species, both annual and perennial. In the bottom footnote of their Table 4, they state that: "(for cultivated annuals) all roots (are) classified as absorptive, assuming that no roots in cultivated annual systems persist through multiple years." This assumption that all crop plant roots are absorptive is at variance with long-standing research with both annual and perennial crop plant species. It is true that, unlike the tree roots upon which the McCormack *et al.* paper is based, there are often few visual clues to the absorptive vs. transport nature of most crop plant roots. The differential visual clues in the long-lived roots of trees are an effective approximation to allow researchers to better refine estimates of actual functional biomass, as well as to help define net primary production (NPP) and carbon sequestration parameters in models.

The McCormack *et al.* review focused on fine roots relative to the "understanding of below-ground contributions to terrestrial biosphere processes". In this context, their discussion and conclusions are very helpful. In the broader sense of fine roots, that is, those of annuals, and first-season perennials vs. those of second- and later-year perennials, the definition of a "fine root," or subdivision of fine roots into categories, is not as easy to deal with. A good example is the work of Jones working with alfalfa (*Medicago sativa*), a perennial crop plant. Jones explained that first-year alfalfa plants and cuttings have a different pattern of fine root development than second- and later-year plants. The primary focus of his paper was on the cambial vs. noncambial roots of second-year plants, and the demonstration that noncambial roots were "ephemeral." This ephemeral nature consisted of the outgrowth of the roots followed by their dieback and the initiation of a new root at or near the base of the previous root. This process was described as occurring multiple times during the season. A similar process in other long-lived perennials is discussed by Paolillo & Zobel and Paolillo & Bassuk. The same process has also been observed in field grown soybean.

Many of the fine roots of second-season plants of alfalfa demonstrated the same morphological differentiation as those described by McCormack *et al.*, but only between first- and second-order roots. The primary difference between the two papers was that McCormack *et al.* describe their absorptive roots as typically living for a year or more. Where this connects or interconnects to the Paolillo group's work needs to be investigated, since they did not demonstrate any longevity time for the "ephemeral" roots their studies found evidence for. Comparing the Paolillo group's results to the McCormack *et al.* descriptions suggests that ephemeral root development may be restricted to much older roots with significant secondary thickening, while the absorptive/transport root pattern is characteristic of the terminal ends of long roots with significant branching. The relative numbers and length of these two types of root have not been determined, therefore also leaving an assessment of relative overall impact undetermined. The extent of the Paolillo ephemeral roots, in quantitative and qualitative terms, needs to be explored if the McCormack *et al.* work is to be refined.

However, this still leaves open the question of the developmental patterns of first-year fine roots on alfalfa and, by extension, other perennials including the long-lived trees. There has been extensive research done on the early development of first-year fine roots for both dicot and monocot

crop species. Unfortunately, the morphological dichotomy described by McCormack *et al.* does not appear to be a consistent rule for first-year fine roots. In first-year fine roots, one to three orders of lateral branching appear to be the rule, although higher orders have been described. In the dicots, the second-order roots rapidly mature, as the lateral roots develop, and develop secondary thickening similar to that described by McCormack *et al.*, allowing the possibility of applying their criteria. In the monocots, however, second (third)-order roots do not develop secondary thickening, but do demonstrate rapid maturation that precludes or reduces absorption. Thus the application of the McCormack *et al.* criteria may be valid for specific cases, while being inadequate or inappropriate for many others. Much research is needed to further compare and contrast these different systems.

To further explore the situation, note that Brown & Ambler demonstrated that only a short region of the tomato (*Solanum lycopersicum*) primary root (between the region of elongation and maturation) was active in the conversion of Fe^{3+} to Fe^{2+} and the resulting uptake and transport, while the fine lateral roots made the conversion and transport along their full length. Zobel *et al.* demonstrated an abrupt reduction in absorption by the parent root when lateral roots began to develop, similar to the pattern described by Brown & Ambler. Zobel *et al.* also demonstrated that in their tomato line (contrary to the observations of Brown & Ambler with a different cultivar of the same species) only the apical 5 cm of the lateral root was highly effective in absorption. The potential changes in function of a given root from absorption to transport over time have been discussed by Waisel & Eshel. It is not clear if this changeover process is also observed in the first- and second-order roots of the long-lived perennial species.

Zobel *et al.* demonstrate species differences in fine root response to nutrient-based changes and Zobel *et al.* demonstrate that the first two orders of roots in chicory (*Chicorum intubus*) may behave in directly opposite ways, and that this can be different among cultivars. Therefore, conclusions from fine-root research data need to be restricted to the species and cultivar actually studied, and not generalized to higher levels without a demonstration of similarity in pattern. Indeed, a careful comparison of the

McCormack *et al.* examples demonstrates that there appear to be differential responses among their long-lived species. As with the fine roots of young and annual plants, generalizations with fine roots of long-lived plants should also be restricted to the cultivar/strain level unless demonstrated to be a more generalized response.

In summary, the functional changes of first-year fine roots as they mature can be visible in dicots showing strong secondary thickening, but they are morphologically more subtle or not discernable for monocots. Thus the morphological duality described in the McCormack *et al.* paper has much more limited application in annual plants, but may be appropriate in some circumstances. If the results described earlier for first-year fine-root development and function in crop plants ultimately translates to second-year and older plants, the morphological dichotomy between absorptive and transport fine roots in trees may serve best as a 'rule of thumb' for long-lived trees when used in a generalized context such as estimates of NPP and potential for carbon sequestration. A resolution of the relationship between Jones and Paolillo's ephemeral roots and the first- through to the fourth-order roots of McCormack *et al.* needs to be explored. Finally, all generalizations must be kept to the cultivar/strain level unless proven to be qualitatively and quantitatively identical in other cultivars, species and higher order plants. (Zobel RW, 2016)

Lesson 37 Plants can gamble, according to study

Imagine you're offered a choice between $800 or a coin toss to win $1000. Heads, you end up with the full $1000; tails, you lose everything. For most of us, it's a no-brainer. We take the $800. But you would likely toss the coin if you were stuck on a desert island with no money, and needed $900 for a flight out. Pea plants, it turns out, make the same decision. When faced with hard times, the species gambles, scientists report. The discovery is the first to show that plants—not just animals—have the ability to switch from being risk-avoiders to risk-takers.

"Like most people, I used to look at plants as passive," says lead author Efrat Dener, a master's student in environmental sciences at

Ben-Gurion University of the Negev in Beersheba, Israel. His group's experiments show "how wrong that view is." Although plants do other things—such as bending toward sunlight and responding to humidity—they haven't been thought of as "dynamic strategists," says Dener's co-author Alex Kacelnik, a behavioral ecologist at the University of Oxford in the United Kingdom. That is, they haven't been shown to be able to respond when times are tough by changing their behavior and taking a chance.

Humans, primates, birds, and social insects take fewer risks when faced with a steady supply of food. But when the supply is uncertain, they switch strategies and take more risks. For instance, in lab experiments, honey bees turn to gambling when they're starving, choosing to sip nectar from a tube that may dispense plentiful amounts or nothing. And dark-eyed juncos (small songbirds) that are cold will ignore a seed dispenser that regularly releases three seeds, and choose one that may give out six or zero.

To find out whether plants do the same, Dener and his colleagues carried out a series of experiments on pea plants (*Pisum sativum*) raised in a greenhouse. The plants were grown with roots split between two pots. Each pot contained the same concentration and type of nutrients. But the level of nutrients in one pot was constant, whereas it varied in the other. After 12 weeks, the scientists measured the plants' root mass and their allocation of roots inside each pot.

They found that the plants varied their distribution of roots depending on the nutrient level in each. In some tests, the plants faced a choice between a pot with a steady supply of high nutrients and one with variable levels. These plants, not surprisingly, were risk-averse, and grew most of their roots in the constant pot.

But plants switched strategies when faced with a choice between a dicey pot with variable levels of nutrients and a pot with constant but low amounts of nutrients—so low, they were below what a plant needs to survive. In this case, the plants, like the person on a desert island, gambled. They sent out more roots in the variable pot, basically tossing a coin to see whether they would get lucky and encounter the nutrients they needed to survive, the scientists report today in *Current Biology*. Thus, normally risk-averse, pea plants become risk-prone when growing in dire conditions.

"To our knowledge, this is the first demonstration" of this kind of risk response in an organism without a nervous system," Kacelnik says. He adds that this doesn't mean the "plants are intelligent" in the way that we think of humans or other animals. But they do have some way of sensing or evaluating the different conditions in the pots, although the scientists do not yet know what this is.

That the scientists have shown risk sensitivity in pea plants—"not really anyone's top candidate for cognitively advanced organism of the year—is a surprise," says David Stephens, a behavioral ecologist at the University of Minnesota, Twin Cities.

But that may be because most people aren't aware of plants' abilities, says James Cahill, a plant ecologist at the University of Alberta, Edmonton, in Canada. "We know plants can process information and have memory." For example, bittersweet nightshade plants can tell whether flea beetles or tortoise beetles are feeding on their leaves, and mount a different chemical defense against each species. The new study "builds on this knowledge. They're showing that risk matters to individual plants. It's a great step and novel contribution to the developing field of plant behavior."

Still, it is a first step, Cahill and Stephens say—but one that is so exciting that both they and the team of researchers are calling for more studies. (Morell V, 2016)

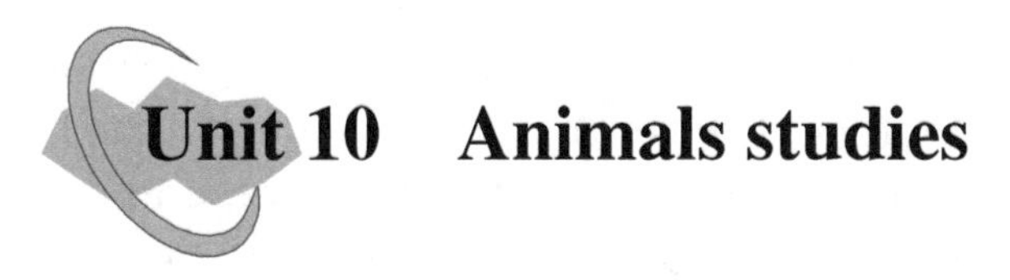

Unit 10 Animals studies

Lesson 38 Marine defaunation: animal loss in the global ocean

Comparing patterns of terrestrial and marine defaunation helps to place human impacts on marine fauna in context and to navigate toward recovery. Defaunation began in earnest tens of thousands of years later in the oceans than it did on land. Although defaunation has been less severe in the oceans than on land, our effects on marine animals are increasing in pace and impact. Humans have caused few complete extinctions in the sea, but we are responsible for many ecological, commercial, and local extinctions. Despite our late start, humans have already powerfully changed virtually all major marine ecosystems.

Humans have profoundly decreased the abundance of both large (e.g., whales) and small (e.g., anchovies) marine fauna. Such declines can generate waves of ecological change that travel both up and down marine food webs and can alter ocean ecosystem functioning. Human harvesters have also been a major force of evolutionary change in the oceans and have reshaped the genetic structure of marine animal populations. Climate change threatens to accelerate marine defaunation over the next century. The high mobility of many marine animals offers some increased, though limited, capacity for marine species to respond to climate stress, but it also exposes many species to increase risk from other stressors. Because humans are intensely reliant on ocean ecosystems for food and other ecosystem services, we are deeply affected by all of these forecasted changes. Three lessons emerge when comparing the marine and terrestrial defaunation experiences: (i) today's low rates of marine extinction may be the prelude to a major extinction pulse, similar to that observed on land during the industrial revolution, as the footprint of human ocean use widens; (ii) effectively slowing ocean defaunation requires both protected areas and careful management of the intervening ocean matrix; and (iii) the terrestrial experience and current trends in ocean use suggest that habitat destruction is likely to become an increasingly dominant threat to ocean wildlife over the next 150 years.

Wildlife populations in the oceans have been badly damaged by human activity. Nevertheless, marine fauna generally are in better condition than terrestrial fauna: Less marine animal extinction has occurred; many geographic ranges have shrunk less; and numerous ocean ecosystems remain more wild than terrestrial ecosystems. Consequently, meaningful rehabilitation of affected marine animal populations remains within the reach of managers. Human dependency on marine wildlife and the linked fate of marine and terrestrial fauna necessitate that we act quickly to slow the advance of marine defaunation. (McCauley DJ, Pinsky ML, Palumbi SR, et al. 2015)

Lesson 39 Tiny DNA tweaks made snakes legless

Sometimes, a genetic tweak can make a really big difference in an animal's appearance. That's what likely happened when the predecessors of modern snakes lost their legs, a process that started some 150 million years ago, two separate groups of scientists have discovered. Although the teams took very different approaches to solve the mystery of how those limbs vanished, both came up with similar results: Mutations in DNA located near a gene key to limb formation keep that gene from ever turning on, they report today.

The new studies have impressed other scientists in the field. In both cases, "there is a correlation between the molecular findings on the one hand and the evolutionary trend of limb reduction and loss on the other," says Michael Richardson, a developmental biologist at Leiden University

in the Netherlands. The findings show "there can be pretty minor changes in the genome that can account for very big specific changes," adds James Hanken, an evolutionary developmental biologist at Harvard University.

Though they are reptiles, almost all snakes are completely missing the limbs typical of most land vertebrates. They didn't start out that way: More than 100 million years ago, snakes had visible legs. And even today pythons and boas have tiny leg bones inside their bodies, suggesting they have vestiges of the molecular pathway for building these appendages.

Scientists got some of their first clues about the genes involved in the development of the serpentine body form in 1999. At the time, Martin Cohn, an evolutionary developmental biologist at the University of Florida in Gainesville, discovered snake embryos had a different pattern of activity of certain genes than other reptiles and that applying a growth factor could make those embryos start to grow limbs. But he lacked the genomic tools to look any deeper. Four years later, Hanken and his colleagues discovered that activity in one of the genes, named after the video game character Sonic the Hedgehog, played a role in leg size in lizards, hinting that it could also be important for snakes. Now, with more ways to monitor gene activity during development—and fully sequenced genomes of various snakes and other reptiles for comparison—Cohn and graduate student Francisca Leal have tracked the genetic activity in embryonic pythons to see why their legs start, but never finish, developing.

They found three DNA deletions in the genetic switch that controls the activity of the *Sonic hedgehog*gene. Situated in front of the gene, this switch, called an enhancer, is a docking site for proteins that control the gene's activity. The deletions made it difficult for certain proteins to land, resulting in only a brief window of gene activity during the development of the python embryo. In snakes with no leg bones, the enhancers have even more deletions, they report, making it likely that the gene never turns on in the first place. "It's taking the old results to a finer level," Hanken says.

Axel Visel, a genomicist at the Lawrence Berkeley National Laboratory in Berkeley, California, also used the python and boa to track down the cause of leglessness. He and his colleagues compared the genomes of those snakes with the genomes of snakes such as vipers and cobras, which evolved more recently and have no leg remnants. They, too, noticed that the same Sonic hedgehog enhancer had many deletions and mutations.

They then showed that a mutated Sonic hedgehog enhancer does indeed affect leg growth by testing its influence on mouse limbs. They replaced the rodent's own version of the enhancer with the snake version using the CRISPR-Cas9 gene-editing technique. They also repeated the experiment, subbing in the same enhancer from a fish and, later, from a human. Even with the fish and human enhancers, the mouse legs developed normally, demonstrating that enhancers from other species still work in the mice. But with the snake enhancer, the legs turned into little nubs, Visel and his colleagues reported in *Cell*. Finally, when researchers put the missing DNA into the snake enhancer and then put the modified enhancer back into mice, their legs grew just fine.

"The Visel paper is a beautiful study; a tour de force that takes functional genomics using CRISPR-Cas9 to a new level," Cohn says. Visel says that Cohn's experiments with snake embryos "independently confirm what we saw in our mouse models. It's very elegant work."

But Cohn, Visel, and Hanken caution that changes in this enhancer are not the whole story behind the evolution of the serpentine form. Recently, for example, other researchers discovered that another enhancer was responsible for snakes' extra ribs and long backbones. And it's possible that the Sonic hedgehog enhancer was not the first step in leg loss, Hanken says. "But it's certainly a major player."

The studies may also help settle a longstanding controversy about fossil snakes, some of which have legs to varying degrees. Paleontologists have long tried to squeeze the limbed fossils onto one branch of the family tree with the limbless ones sprouting off from that branch, something that would be expected had limbs been lost only once. But if it didn't take much to lose legs, then it probably didn't take much to re-evolve them. "It could explain the possible reappearance of limbs in some extinct snake

lineages," Richardson says. For centuries, evolutionary biologists have argued that organisms cannot re-evolve lost features. But Hanken says this work "shows that, developmentally speaking, it's not that farfetched." (Pennisi, 2016)

Lesson 40 How Earth's oldest animals were fossilized

The fossils are among the strangest ever found: a corkscrew-shaped tube, an eight-armed spiral, and a mysterious ropelike creature that might have engaged in the oldest known sexual reproduction among animals. They are Earth's oldest complex organisms, dating back to 571 million years ago, and found on every continent except Antarctica. Their bizarre forms defy classification; some have been described alternately as jellyfish or worms, algae or fungi. But scientists have for years been chasing an even bigger mystery about the so-called Ediacara biota: How could these mostly soft-bodied animals be preserved in rock? Now, one team of scientists has an answer. Their research suggests that in the ancient oceans, silica—the primary compound in quartz—precipitated out of the seawater, then covered and entombed the organisms before they decayed.

This research will change our way of thinking about Ediacara-type preservation," says James Schiffbauer, a paleontologist at the University of Missouri in Columbia, who was not involved in the new study. He adds that the process might not be as straightforward as scientists thought.

Most fossils exist thanks to how they were buried plus the makeup of their original tissues. Bones and shells from hard-bodied creatures like dinosaurs and oysters preserve more easily than soft tissues, which decay rapidly after death. That means that most of the fossil record is biased in favor of creatures with hard components. "One of the big questions that we have in really all of paleontology ... is how accurately can we read the fossil record as the history of life?" Schiffbauer says.

Before the appearance of the Ediacara biota, named for the Ediacara Hills in South Australia where scientists first found these fossils, all known

life on Earth was microscopic. That's because scientists hadn't found any evidence of complex life until the "geologically abrupt" entrance of the Ediacaran fossils, says Yale University paleontologist Lidya Tarhan, lead author on the new study, which was published in Geology. But is this sudden explosion of the fossil record just a preservation bias or is it a sign of a massive environmental trigger for the biota's emergence? Finding out how the group became fossils "is one of the most important steps in resolving what these organisms are and where they fall in our sense of the evolution of complex life," she says.

So Tarhan and her team set out to find the answer. They knew the animals lived in shallow waters on the sea floor, and that sand stirred by storms would sometimes cover the organisms. The leading theory for their preservation was that these sand grains molded around dead bodies, and the mold continued to exist long after the bodies decayed. For that to happen, "you have to cement those grains, and you have to do it early," Tarhan says. Previous work hadn't addressed how that cementing could have happened. But Tarhan's team had a theory: Researchers knew the Ediacaran oceans contained far more dissolved silica than modern ones, in part because creatures that soak up silica, like sponges, were rare. So silica was the perfect candidate for a prehistoric glue.

To test their hypothesis, the team took fossils from the South Australian outback and sawed them into slivers of rock so thin that light, passing through them under a microscope, illuminated the ancient grains. "The grains are pretty much floating in what looks like a sea of cement, and they're not very compacted," Tarhan says. Her team confirmed that the "sea" was indeed silica. And because the grains weren't compacted, they must have been loose as the silica cement formed around them. Finally, the team concluded that the silica-based cements were not chemically identical to the silica found in the quartz sand grains, leaving them with only one source for the cement: seawater.

Because this style of siliceous fossilization extends long before and after the Ediacaran, the biota's appearance—and disappearance—was not just an accident of the fossil record, Tarhan says. Instead, they must

represent the group's actual evolutionary beginning as well its ultimate extinction. "It makes a lot of sense," says Shuhai Xiao, a geobiologist at Virginia Polytechnic Institute and State University in Blacksburg, who was also not involved in the study. "The next step is to take this model somewhere else, and to test it to see if it works" at other Ediacaran sites around the world. (Joel, 2016)

Lesson 41 Why do zebras have stripes?

Zebras in warmer climates sport more stripes, perhaps to keep them cool or healthy. A leopard may not be able to change its spots, but some zebras change their stripes. Zebras in warmer places have more stripes, a new study shows, which might help answer an age-old question: Why stripes?

The answer probably comes down to keeping zebras cool and fending off disease-causing insects that are more common in hotter climates, researchers reported in the journal *Royal Society Open Science.*

All three species of zebra have bold black and white stripes that stand out among more drab-looking African grazers, like buffalo and antelope, especially against a plain savanna background. And standing out would seem to make a zebra more likely to become a lion's lunch.

This "stripe riddle" has puzzled scientists, including Darwin, for over a century. There are five main hypotheses for why zebras have the stripes: to repel insects, to provide camouflage through some optical illusion, to confuse predators, to reduce body temperature, or to help the animals recognize each other.

A new analysis of the plains zebra—the most common species, which ranges from Ethiopia to South Africa—doesn't tease out one theory as the definitive winner. But it does show that temperature is the factor most strongly linked to striping: More specifically, the warmer it is, the more stripes on the zebra.

Of every stripe

Brenda Larison, a biologist at the University of California, Los Angeles, and colleagues visited 16 zebra populations throughout Africa and studied their stripe patterns, in a project supported by National Geographic Society's Committee for Research and Exploration.

The team then measured 29 different environmental factors—such as soil moisture, rainfall, prevalence of disease—carrying tsetse flies, and distribution of lions and plugged them into a computer model to see which ones were related to differences in stripe patterns across the zebra's range.

The two factors that mattered most, said Larison, were how consistent the temperature was in a particular area and the average temperature during the coldest part of the year.

The researchers then went a step further, using the two temperature variables to predict the striping patterns of zebra populations not included in the study.

"We were able to show that we could predict it with significant accuracy," Larison said.

Cooling effect?

Why temperature affects striping is another question, she said, but there are two possible reasons.

One is the "cooling eddy" theory. When air hits a zebra, the currents are stronger and faster over the black parts (since black absorbs more heat than white) and slower over the white. At the juncture of these two opposing airflows, little eddies of air may swirl and serve to cool a zebra's skin. For instance, Larison said, there's evidence that heavily striped zebras have 5.4-degree Fahrenheit (3 degrees Celsius) lower skin temperatures than other non-striped mammals in the same area.

The other idea holds that more stripes may be a barrier against disease, since disease-carrying biting flies, like horseflies, tend to like it hot. Experiments in the field have shown that biting flies don't like landing on striped surfaces.

Tim Caro, a biologist at the University of California, Davis, supports the disease theory. "We're getting a lot of similarities in our findings," said Caro, whose own research showed that striping is linked to repelling biting flies.

"Diseases carried by horseflies are really nasty," he said. "They can hold a lot of diseases like equine influenza, and it's possible that those diseases are going to be more of a problem under warmer, wetter conditions." (Dell'Amore, 2015)

Lesson 42 Tiny microbe turns tropical butterfly into male killer

A scientist from the University of Exeter has helped to identify a male-killing microbe in a tropical butterfly called the African Queen, which leads to the death of all sons when a mother is infected.

In most of Africa this microbe, called *Spiroplasma*, infects African Queen butterflies but has no effect on their offspring. However, in a narrow zone around Nairobi in Kenya, where two sub species of butterfly live and breed, the scientists noted that the microbe infection caused all their sons to die. In fact, the male eggs never hatch and are often consumed by their hungry sisters.

The authors of the paper, published in the journal *Proceedings of the Royal Society B*, believe that the phenomenon, which takes place where two sub species meet, is the first step in the transition of the two sub-species into two true, non-interbreeding, species.

Professor Richard Ffrench-Constant, from the Centre for Ecology and Conservation at the University of Exeter's Cornwall campus, and a team of British, Kenyan and German scientists, have found that the chromosomes of the females in which male-killing occurs have changed dramatically with a non-sex chromosome fusing with a sex chromosome to form a new chromosome called "neo W".

Professor Ffrench-Constant, Professor of Molecular Natural

History, said: "We tend to think of new species coming about due to environmental changes but here it's clearly the microbe that is driving these two subspecies apart."

"Whilst we don't understand the precise molecular mechanisms behind this chromosomal merger, this means that no males are made in the hybrid zone, and that mating success in the zone is effectively zero, thereby creating a barrier with a new species on either side."

This paper represents the culmination of 13 years of field work in which the sex and colour pattern of butterflies around Nairobi was painstakingly recorded by a team led by Dr. Ian Gordon based in Nairobi. The breakthrough came when female butterflies from the all-female zone were sent to Germany to have their chromosomes examined and where Professor Walther Traut from the University of Lübbek discovered that two of the chromosomes had fused.

Dr. David Smith, formerly from the Natural History Museum at Eton College, first author on the paper, said: "The neo-W effectively acts as a genetic sink for all males, and butterfly populations around Nairobi are nearly all female. Our results demonstrate how a complex interplay between sex, colour pattern, male-killing and chromosomes has set up a genetic 'sink' that keeps two subspecies apart."

Professor Walther Traut, from the University of Lübbek, said: "This is like a smoking gun for the way in which species become distinct. It is rare that we can find the molecular basis for how species develop."

Professor Ffrench-Constant added: "It appears that the butterfly's susceptibility to the male-killing microbe is driving the separation of the two butterflies into two true species. These tiny microbes are therefore having a major effect on sex and death in this fascinating butterfly." (University of Exeter, 2016)

Lesson 43 Dogs recognize dog and human emotions

The recognition of emotional expressions allows animals to evaluate the social intentions and motivations of others. This provides crucial information about how to behave in different situations involving the establishment and maintenance of long-term relationships. Therefore, reading the emotions of others has enormous adaptive value. The ability to recognize and respond appropriately to these cues has biological fitness benefits for both signaller and the receiver.

During social interactions, individuals use a range of sensory modalities, such as visual and auditory cues, to express emotion with characteristic changes in both face and vocalization, which together produce a more robust percept. Although facial expressions are recognized as a primary channel for the transmission of affective information in a range of species, the perception of emotion through cross-modal sensory integration enables faster, more accurate and more reliable recognition. Cross-modal integration of emotional cues has been observed in some primate species with conspecific stimuli, such as matching a specific facial expression with the corresponding vocalization or call. However, there is currently no evidence of emotional recognition of heterospecifics in non-human animals. Understanding heterospecific emotions is of particular importance for animals such as domestic dogs, who live most of their lives in mixed species groups and have developed mechanisms to interact with humans. Some work has shown cross-modal capacity in dogs relating to the perception of specific activities (e.g., food-guarding) or individual features (e.g., body size), yet it remains unclear whether this ability extends to the processing of emotional cues, which inform individuals about the internal state of others.

Dogs can discriminate human facial expressions and emotional sounds. However, there is still no evidence of multimodal emotional integration and these results relating to discrimination could be explained through simple associative processes. They do not demonstrate emotional recognition, which requires the demonstration of categorization rather than differentiation. The integration of congruent signals across sensory inputs

requires internal categorical representation and so provides a means to demonstrate the representation of emotion.

Albuquerque et al. (2016) used a cross-modal preferential looking paradigm without familiarization phase to test the hypothesis that dogs can extract and integrate emotional information from visual (facial) and auditory (vocal) inputs. If dogs can cross-modally recognize emotions, they should look longer at facial expressions matching the emotional valence of simultaneously presented vocalizations, as demonstrated by other mammals. Owing to previous findings of valence, side, sex and species biases in perception studies, they also investigated whether these four main factors would influence the dogs' response. Their results indicated that dogs with either human or dog faces with different emotional valences (happy/playful versus angry/aggressive) paired with a single vocalization from the same individual with either a positive or negative valence or Brownian noise. Dogs looked significantly longer at the face whose expression was congruent to the valence of vocalization, for both conspecifics and heterospecifics, an ability previously known only in humans. These results demonstrate that dogs can extract and integrate bimodal sensory emotional information, and discriminate between positive and negative emotions from both humans and dogs. (Albuquerque N, Guok K, Wilkinson A, et al. 2016)

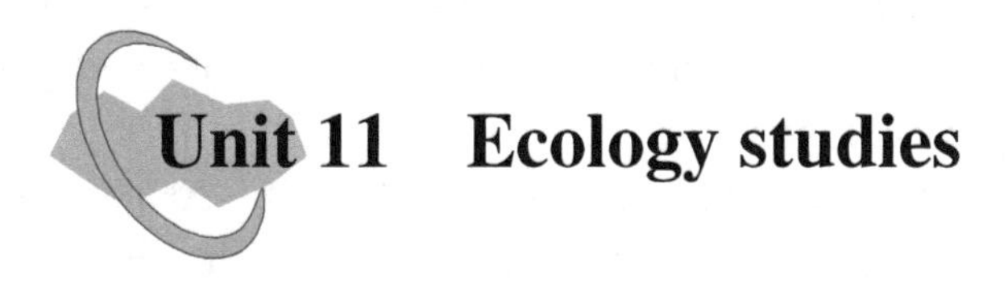

Unit 11 Ecology studies

Lesson 44 The next century of ecology

The science of ecology is about relationships—among organisms and habitats, on all scales—and how they provide information that helps us better understand our world. In the past 100 years, the field has moved from observations to experiments to forecasting. Next week, the Ecological Society of America (ESA), the world's largest ecological society, celebrates its centennial in Baltimore, an opportunity to reflect on the field's past and future. The gathering of international scientists, policy-makers, and students will not only explore the knowledge in hand, but consider what else is needed to chart a course over the next century in which humanity sustains and even improves the relationships that underpin life on Earth.

Characterized by a focus on interactions, from genes to global scales, and between living and nonliving components of ecosystems, basic ecological research has spawned important paradigm changes over the past 100 years. For example, we have learned that a simple graphical model of biogeography can explain species distribution patterns at many spatial scales. Another major change has been the development of our understanding of succession after disturbances, from major forest fires to the effects of antibiotics on intestinal microbial communities. As ecological science becomes more interdisciplinary, shifts in thinking and unexpected impacts will continue. Early ecologists who thought about principles governing plant and animal communities never imagined that their ideas would provide the foundation for understanding the human microbiome, affecting our nutrition, immune system, and even psychological state. The new field of synthetic ecology, in which ecologists and medical professionals design beneficial microbial communities, has its origins in century-old ecological field studies. These examples foretell how the roles of ecologists and the applications of ecological principles are likely to change in the next century, and why medical students and practitioners need to understand ecology.

The good news is that ecology's role in society has grown dramatically over the past century. Basic research on organism and environment interactions has had far-reaching impacts on legislation. In the United States, this knowledge has contributed the scientific basis for the landmark Endangered Species Act (1973) and Clean Water Act (1972). In this vein, in the 1990s, the ESA's Sustainable Biosphere Initiative encouraged the world's ecologists to identify major environmental challenges. Today, its Earth Stewardship Initiative frames a commitment by ecologists to make their science relevant to society through activities ranging from practical demonstrations to communication campaigns directed at other communities, including communities of faith. ESA's meeting in Baltimore will highlight ongoing studies of the city's urban ecology as a model to create more livable and sustainable cities, linking environmental stewardship to design and planning.

The international community is also placing more emphasis on ecology, as demonstrated by the new Intergovernmental Platform on Biodiversity and Ecosystem Services. Its first assessment, focused on pollinators, and reports will provide actionable science-based recommendations that should catalyze better incorporation of ecological science into management and legislation. The new U.S. National Ecological Observatory Network should elucidate the importance of the biosphere in Earth system dynamics and its governing role in climate, a resource for all countries. With newly developed tools, analytical methods, and models to forecast the future of the world's environment, ecologists can inform policy-makers and political leaders about how to prevent, mitigate, or adapt to environmental change.

From the microbes inhabiting the earth beneath our feet to environments of the universe unknown to us now, the next 100 years of ecological discoveries will influence our lives. We enter a time when society is armed with the scientific knowledge and ability to make responsible decisions. (Inouye DW, 2015)

Lesson 45 Ninety-nine percent of the ocean's plastic is missing

Millions of tons—That's how much plastic should be floating in the world's oceans, given our ubiquitous use of the stuff. But a new study finds that 99% of this plastic is missing. One disturbing possibility: Fish are eating it.

If that's the case, "there is potential for this plastic to enter the global ocean food web," says Carlos Duarte, an oceanographer at the University of Western Australia, Crawley. "And we are part of this food web."

Humans produce almost 300 million tons of plastic each year. Most of this ends up in landfills or waste pits, but a 1970s National Academy of Sciences study estimated that 0.1% of all plastic washes into the oceans from land, carried by rivers, floods, or storms, or dumped by maritime vessels. Some of this material becomes trapped in Arctic ice and some, landing on beaches, can even turn into rocks made of plastic. But the vast majority should still be floating out there in the sea, trapped in midocean gyres-large eddies in the center of oceans, like the Great Pacific Garbage Patch.

To figure out how much refuse is floating in those garbage patches, four ships of the Malaspina expedition, a global research project studying the oceans, fished for plastic across all five major ocean gyres in 2010 and 2011. After months of trailing fine mesh nets around the world, the vessels came up light-by a lot. Instead of the millions of tons scientists had expected, the researchers calculated the global load of ocean plastic to be about only 40,000 tons at the most, the researchers report online today in the *Proceedings of the National Academy of Sciences*. "We can't account for 99% of the plastic that we have in the ocean," says Duarte, the team's leader.

He suspects that a lot of the missing plastic has been eaten by marine animals. When plastic is floating out on the open ocean, waves and radiation from the sun can fragment it into smaller and smaller particles, until it gets so small it begins to look like fish food—especially to small lanternfish, a widespread small marine fish known to ingest plastic.

"Yes, animals are eating it," says oceanographer Peter Davison of the Farallon Institute for Advanced Ecosystem Research in Petaluma, California, who was not involved in the study. "That much is indisputable." But, he says, it's hard to know at this time what the biological consequences are. Toxic ocean pollutants like DDT, PCBs, or mercury cling to the surface of plastics, causing them to "suck up all the pollutants in the water and concentrate them." When animals eat the plastic, that poison could be going into the fish and traveling up the food chain to market species like tuna or swordfish. Or, Davison says, toxins in the fish "may dissolve back into the water … or for all we know they're puking or pooping it out, and there's no long-term damage. We don't know."

It's impossible to know how much the animals are eating, says Kara Law, a physical oceanographer at the Sea Education Association in Woods Hole, Massachusetts, who was not involved in the work. The estimated amount of plastic entering the ocean that the study uses is almost half a century old, and "we're desperately in need of a better estimate of how much plastic is entering the ocean annually."

What's more, both Davison and Law say there are a number of other potential places the plastic could be ending up. It could be washing ashore, and a lot of it could be degrading into pieces too small to be detected. Another possibility is that organisms sticking to and growing on the plastic are dragging the junk beneath the ocean's surface, either suspending it in the water column or sinking it all the way to the sea floor. Microbes may even be eating the stuff.

Best-case scenario for the fate of the missing plastic? It's sinking from the weight of organisms sticking to it or in animal feces and getting buried on the ocean floor, Law says. "I don't think we can conceive of the worst-case scenario, quite frankly. We really don't know what this plastic is doing." (Chen A, 2014)

Lesson 46 Earth's lakes are warming faster than its air

The world's lakes are warming faster than both the oceans and the air

around them, a global survey of hundreds of lakes shows. The rapid temperature rise could cause widespread damage to lake ecosystems, say scientists who presented the findings today at the American Geophysical Union meeting here. The global effects could be even more serious, because higher lake temperatures could trigger the conversion of billions of tons of carbon stored in lake sediments to methane and carbon dioxide (CO_2), in a feedback effect that could accelerate global warming.

The temperature increase—a summertime warming of about a third of a degree per decade over 25 years—is "pretty modest," says lake biologist Peter Leavitt of the University of Regina in Canada, who did not participate in the study. "But you don't need 2°- to 3°- increases in lake temperatures to have profound impacts."

Many studies have shown that individual lakes are warming up in the summer. (Scientists rarely track lake temperatures in winter, when ice makes measurements more challenging.) The new study encompassed 235 lakes around the world, combining measurements by hand, which reach the depths of lakes, with satellite readings, which provide global coverage of lake surfaces. Over the study's time period, between 1985 and 2009, a few lakes cooled whereas others warmed sharply, but the average warming of 0.34°C per decade was more than twice the 0.12°C per decade measured in the oceans over a similar period.

That the oceans lag isn't unexpected, given their enormous mass, says Catherine O'Reilly, lead author on the study, which involved 64 scientists collecting data on six continents. But many lakes are warming even faster than surface air temperatures, which warmed by 0.25°C per decade between 1979 and 2012. "I would have expected that, on average, lakes would be warming more slowly than air," says O'Reilly, a freshwater ecologist at Illinois State University in Normal.

A shorter ice season because of warmer winters may help explain why. "Normally, ice is a good insulator protecting lakes from atmospheric heating," Leavitt says. With ice melting earlier, lake water is exposed to warm spring air for longer. That probably explains why lakes that normally freeze in winter are warming by 0.48°C per decade, about twice as fast as lakes that don't freeze.

Another factor that may be accelerating the lake warming is a decline in cloudiness in some temperate areas, due at least in part to climate change. Clearer air allows more sunlight to strike lakes' surfaces. And as lakes soak up more heat from the sun and the air, their waters become more stratified, with the less dense warm water floating on top of more dense cold water. The stratification prevents deep, cold waters from mixing into surface layers and cooling them in summer.

The rapid summertime warming bodes ill for lake species. Freshwater fish that like the cold, such as lake trout, could suffer. So could species that rely on increasingly threatened lake ice. The Baikal seal in Russia's Lake Baikal gives birth on the ice, points out biologist Stephanie Hampton of Washington State University, Pullman. Adds O'Reilly: "Large changes in our lakes are not only unavoidable, but are probably already happening."

Meanwhile, in warmer places, says Leavitt, "strong stratification and warm surface waters are the recipe for blooms of noxious and possibly toxic cyanobacteria, particularly in regions where agriculture and urbanization have fertilized the lakes and estuaries."

Warming lakes may have global implications as well, say the researchers whose study will be published today in *Geophysical Research Letters*. As aquatic organisms die, their carbon-rich remains fall into the water column, where they can be stored in sediments or broken down by microorganisms into gases. "Lakes are already massive furnaces for processing terrestrial organic matter" and creating greenhouse gasses, Leavitt says. "Warming these regions further is likely to increase their role in combusting carbon to CO_2."

Kintisch says their findings point to the need to update global climate models so they better predict lake warming. They'd also like to see better remote sensing techniques that can measure the temperature of smaller lakes, missed by current satellites. (Kintisch E, 2015)

Lesson 47 Strong invaders are strong defenders—implications for the resistance of invaded communities

The spread of non-native species is likely to continue and possibly

accelerate in the future as a consequence of increasing transportation of goods and people. This trend is worrisome because invasions by non-native species can have considerable negative effects on ecosystems and the services that they deliver. However, invasion by non-native species also provides opportunities for the study of community assembly. An important question that has been raised in the context of invasion biology is: Will the establishment of new species make communities more vulnerable to further invasions or will they become increasingly resistant to additional invaders? Insights into answering this question can be obtained by studying how the invasion success of species is correlated with their contribution to resistance.

The outcomes of species introductions are to a large extent determined by interactions between introduced and resident species. Non-native species that are efficient predators, strong resource competitors or apparent competitors, or are resistant against natural enemies, are expected to be successful in their interactions with resident species. Conversely, these attributes should make resident species successful in their interactions with non-native species. Assessing the strengths of species interactions is, however, very time consuming, and it is in principle much easier to assess the outcome of introductions and subsequently draw inferences about species interactions from performance measures. Four such measures can be derived for each species in their respective roles as invader vs. resident: (1) invasion success—the ability to establish successfully, (2) impact—the ability to cause density reductions or local extinction of residents, (3) defence capacity—the contribution to resistance of the resident community to invasion and (4) persistence—the ability of a resident, either native or non-native, to persist and maintain high population densities after the establishment of invaders.

In the study, Henriksson et al. (2016) utilized a unique dataset on 1157 well-documented, human introductions (about half of which failed) of 26 different fish species into 821 Swedish lakes to search for correlation patterns among performance traits. Specifically they searched for correlations across species in the different performance traits, invasion success, impact, defence capacity and persistence, and used this information to draw inferences on the development of biotic resistance in

Swedish lake fish communities. They found that the observed correlation patterns are consistent with a hierarchical community structure with some species performing well both as invaders and residents, and others performing poorly on both counts. Thus, they conclude that future invasions will make these communities more resistant. (Henriksson A, Wardle DA, Trygg J, et al. 2016)

Lesson 48 Global warming favours light-coloured insects in Europe

During the last decades, considerable interest has arisen in predicting the distribution of species, assemblages of species and characteristics of assemblages under various scenarios of climate change. Most of these predictions, however, are based on phenomenological models and provide little mechanistic understanding of the underlying processes. A first step towards understanding the impact of climate change is to consider traits of species that underpin the relationship between species and climate. This approach has been very successful for plants. For instance, leaf size as well as leaf shape have been used to reconstruct temperature and precipitation of paleoclimates. However, the relationships between traits of animals and climate are more difficult to identify. One trait—the body size of endotherms—may be related to climate. Bergmann's rule predicts a decrease in their body size within species and across closely related species with increasing temperature. In line with this ecogeographic rule, a recent study found that the body size of some mammals decreases with increasing temperature. Although many physiological and local-scale studies on insects have demonstrated associations between traits and climate, to what extent such mechanistic links may impact distribution patterns of insect species at a biogeographical scale remains largely unknown.

In ectotherms, the appearance of the body surface and in particular its colour value are involved in thermoregulation. Dark-coloured ectotherms are able to increase their body temperatures above ambient air temperature more effectively than light-coloured ectotherms, and therefore have an

advantage in cool climates (thermal melanism hypothesis). Most insects—by far the most species-rich lineage of ectotherms—need to reach body temperatures above ambient temperature for flying, foraging or mating. However, being dark is only advantageous in cool climates. In areas with high temperatures and insolation, insects need to protect themselves against overheating. At high temperatures, ectothermic species with light colouration can be active for a longer period than species with dark colouration, and may be able to use a broader thermal range of habitats. Overall, heat avoidance and heat gain constraints lead to the prediction that insect assemblages should be dominated by dark-coloured species in cool climates and by light-coloured species in warm climates. The first hint that this holds even at a biogeographical scale was provided by Rapoport almost 50 years ago, when he used a rough estimate of springtail colour value to show a positive correlation of the percentage of dark-coloured species in species assemblages with latitude and altitude.

In this study experiment by Zeuss et al., they combine recent digital image analysis and phylogenetic statistics to demonstrate that colour lightness of insects is consistently correlated to the thermal environment across Europe. They furthermore show that assemblages of dragonflies became on average lighter-coloured during the last century, which we attribute to global warming. (Zeuss D, Brandl R, Brandle M, et al. 2014)

Lesson 49 Carbon dioxide supersaturation promotes primary production in lakes

Most of the lakes in the world are unproductive with inputs of organic carbon dominated by terrestrial sources. Terrestrial organic carbon is used as a carbon and energy source by lake water bacterioplankton with two major consequences; increased bacterial production and biomass and increased production of CO_2 via bacterial respiration. Hence, bacterioplankton become an important base for production in lake food webs and many lakes get supersaturated with respect to CO_2. The CO_2-supersaturation and bacterial respiration are proportional to the terrestrial export of dissolved organic carbon (DOC), and pCO_2 can often reach values that are a magnitude or more above the concentrations at equilibrium with the atmosphere. Supersaturated lakes are, therefore, net sources of CO_2 to the atmosphere, and a significant part of global carbon cycling.

In contrast to the large recent interest in lakes as CO_2 sources to the atmosphere, there have been only a moderate number of efforts to explore possible ecological effects of CO_2-supersaturation in unproductive lakes. One such effect is the possible stimulation of phytoplankton primary production (PP). Studies of marine, and freshwater phytoplankton, thus, show that algal photosynthesis rates can be enhanced by increased CO_2 concentrations both at undersaturation (nutrient rich and productive systems) and supersaturation (nutrient poor and unproductive systems). These observations, which indicate a carbon limitation of phytoplankton have, however, not inspired any systematic analysis of the role of CO_2-supersaturation for lake primary productivity. A possible reason is that, ever since the "carbon-phosphorus controversy" in the early 1970s, there has been a consensus among freshwater ecologists that CO_2 plays a minor role as a limiting factor for PP compared with phosphorus (P) and nitrogen (N), except for short periods in highly productive lakes. It is also counterintuitive to ascribe CO_2 the role of a limiting factor in unproductive lakes where its concentration is high and the concentrations of P and N are low.

By experimental manipulation of the CO_2 concentration in supersaturated boreal lakes. Jansson et al. demonstrated that phytoplankton primary production was up to 10 times higher in supersaturated lake water in comparison with water with CO_2 at equilibrium concentrations and that CO_2, together with nutrients, explained most of the variation in pelagic primary production and phytoplankton biomass over a wide variety of unproductive lakes. These results suggested that phytoplankton could be co-limited by CO_2 and nutrients in unproductive lakes. As import of terrestrial organic carbon and its subsequent microbial mineralisation in lakes was a driving force of CO_2-supersaturation. These results also indicated that lake productivity and carbon cycling may respond to variations in terrestrial organic carbon export (e.g., caused by land use or climate change) in ways not described before. (Jansson M, Karlsson J, Jonsson A, et al. 2012)

Part III Scientific Writing and Publishing in English

英语科技论文写作与发表

Unit 12　Common problems and corresponding suggestions in document indexing

Lesson 50　Common problems in document indexing

一篇好的论文，必须通过查阅大量的书籍和文献，在借鉴别人学术成果的基础上实现再创造，形成自己独立的思考能力和自己独到的见解。20 世纪 90 年代中期以来，计算机网络技术在我国发展迅猛。相应地，文献检索与利用的对象也发生了巨大的转变，时至今日，电子文献已经取代传统的纸质文献，成为文献检索与利用的主角。今天的大学生基本上都能熟练应用计算机（杨冬艳，2009），但如何高效地检索到所需文献是当代大学生面临的问题。下面就针对当前大学生在文献检索中遇到的主要问题总结如下。

一、所知检索途径少

根据上课调查发现，百度是大学生最喜爱的检索途径之一。相比之下，学校购买的数据库则很少有人用。在众多数据库中，大学生在检索文献时，更喜欢使用有限的几个中文数据库，如中国知网、维普和万方。然而，学校购买的英文数据库却很少有学生关注。但是，外文文献是大学生学习中不可忽视的宝贵资源，对于生物学专业的大学生来讲尤其如此。为了方便学生查阅外文资料，下面就国外主流数据库的特点及使用方法进行简单介绍。希望大家能够借助这些工具来提高自己。

外文文献数据库有多种分类，其中从所收录文献信息的使用方式的角度可分为两类：第一类是收录文献全文的数据库，以ScienceDirect、Springer为代表；第二类是收录摘要、文献来源和文献引证关系的数据库，以所谓的三大索引数据库（科学引文索引）SCI、（工程索引）EI、（科技会议录索引）ISTP 为代表（三大索引数据库统一于 ISI web of knowledge）。以下对这两类数据库逐一介绍。

（一）收录全文的数据库

1. ScienceDirect 数据库

Elsevier 是荷兰一家全球著名的学术期刊出版商，ScienceDirect 是荷兰 Elsevier 公司出版的全球最全面的全文文献数据库，涵盖了几乎所有学科领域。每年出版大量的学术图书和期刊，大部分期刊被 SCI、SSCI、EI 收录，是世界上公认的高品位学术期刊。ScienceDirect 得到了 70 多个国家的认可，是目前中国国内使用率最高、下载量最多的科学数据库。

2. Springer Link 数据库

Springer 电子期刊数据库是德国施普林格（Springer-Verlag）世界著名科技出版集团的产品，通过 Springer Link 系统提供学术期刊及电子图书的在线服务。目前 Springer Link 所提供的全文电子期刊按学科分为 11 个在线图书馆：（1）生命科学（life sciences）含 105 种期刊；（2）医学（medicine）含 168 种期刊；（3）数学（mathematics）含 77 种期刊；（4）化学（chemical sciences）含 36 种期刊；（5）计算机科学（computer sciences）含 45 种期刊；（6）经济（economics）含 31 种期刊；（7）法律（law）含 5 种期刊；（8）工程学（engineering）含 52 种期刊；（9）环境科学（environmental sciences）含 34 种期刊；（10）地球科学（geosciences）含 50 种期刊；（11）物理学与天文学（physics and astronomy）含 61 种期刊。该系统收录 1996 年至今的期刊，

1996 年以前的期刊将逐步开通。Springer Link 中的大多数全文电子期刊是国际重要期刊，它是科研人员的重要信息源。

3. Wiley Online Library 数据库

作为全球最大、最全面的经同行评审的科学、技术、医学和学术研究的在线多学科资源平台之一，“Wiley Online Library”覆盖了生命科学、健康科学、自然科学、社会与人文科学等全面的学科领域。它收录了来自 1500 余种期刊、10 000 多本在线图书及数百种多卷册的参考工具书、丛书系列、手册和辞典、实验室指南和数据库的 400 多万篇文章，并提供在线阅读。

4. ProQuest 数据库

ProQuest 数据库收录了 1861 年以来全世界 1000 多所著名大学理工科 160 万博士、硕士学位论文的摘要及索引，学科覆盖了数学、物理、化学、农业、生物、商业、经济、工程和计算机科学等，是学术研究中十分重要的参考信息源。

5. NCBI 数据库

美国国立生物技术信息中心（National Center for Biotechnology Information），即我们所熟知的 NCBI，它是由美国国立卫生研究院（NIH）于 1988 年创办。创办 NCBI 的初衷是为了给分子生物学家提供一个信息储存和处理的系统。除了建有 GenBank 核酸序列数据库（该数据库的数据资源来自全球几大 DNA 数据库，其中包括日本 DNA 数据库 DDBJ、欧洲分子生物学实验室数据库 EMBL 及其他几个知名科研机构）之外，NCBI 还可以提供众多功能强大的数据检索与分析工具。目前，NCBI 提供的资源有 Entrez、Entrez Programming Utilities、My NCBI、PubMed、PubMed Central、Entrez Gene、NCBI Taxonomy Browser、BLAST、BLAST Link （Blink）、Electronic PCR 等共计 36 种功能，而且都可以在 NCBI 的主页 www.ncbi.nlm.nih.gov 上找到相应链接，其中多半是由 BLAST 功能发展而来的。

（二）收录摘要、文献来源和文献引证关系的数据库

1. SCI 数据库

SCI（Science Citation Index）即科学引文索引，是由美国科学信息研究所（ISI）1961 年创办出版的引文数据库，是自然科学领域基础理论学科方面的重要期刊文摘索引数据库。SCI（科学引文索引）、EI（工程索引）、ISTP（科技会议录索引）是世界著名的三大科技文献检索系统，是国际公认的进行科学统计与科学评价的主要检索工具，其中以 SCI 最为重要。SCI 所收录期刊的内容主要涉及数学、物理、化学、农业、林业、医学、生物等基础科学研究领域，选用刊物来源于 40 多个国家，文字涉及 50 多种，其中主要的国家有美国、英国、荷兰、德国、俄罗斯、法国、日本、加拿大等，也收录部分中国刊物。

它不仅是一部重要的检索工具书，而且也是科学研究成果评价的一项重要依据。它已成为目前国际上最具权威性的、用于基础研究和应用基础研究成果的重要评价体系。它是评价一个国家、一个科学研究机构、一所高等学校、一本期刊，乃至一个研究人员学术水平的重要指标之一。

2. EI 数据库

EI（Engineering Index）即工程索引，是世界著名的检索工具，由美国工程信息公司（Engineering Information Inc.）编辑出版发行，该公司始建于 1884 年，是世界上最大的工程信息提供者之一，早期出版印刷版、缩微版等信息产品，1969 年开始提供 EI Compendex 数据库服务。EI 以收录工程技术领域的文献全面且水平高为特点。EI 收录 5000 多种工程类期刊论文、会议论文和科技报告。收录范围包括核技术、生物工程、运输、化学和工艺、光学、农业和食品、计算机与数据处理、应用物理、电子与通信、材料、石油、航空和汽车工程等学科领域。

3. ISTP 数据库

ISTP（Index to Scientific & Technical Proceedings）即科技会议录索引，创刊于 1978 年，由美国科学情报研究所编辑出版。该索引收录生命科学、物理与化学科学、农业、生物和环境科学、工程技术和应用科学等学科的会议文献，包括一般性会议、座谈会、研究会、讨论会、发表会等。ISTP 收录论文的多少与科技人员参加的重要国际学术会议多少或提交、发表论文的多少有关。中国科技人员在国外举办的国际会议上发表的论文占被收录论文总数的 64.44%。

二、缺乏必要的检索技巧

掌握文献数据库检索方法与检索技巧，能提高检索效率和准确度，能在既定的检索范围内最大限度地采集检索者所需的信息。笔者曾遇到多名大学生在检索文献方面存在困惑——他们在使用数据库检索论文时输入自己想要的关键词，结果出来几千篇，甚至几万篇相关论文；他们不能够找到和自己直接相关的文献，有的即使找到了和自己相关的论文，但因为学校没有购买所包含该期刊的数据库而看不到全文。通过调查发现，大学生在文献检索时缺乏一些必要的技巧和方法。下面将介绍大学生在文献检索中遇到的以下几种情况所需要的一些基本技巧和方法。

1. 以系统默认的检索项检索

大学生在进行文献检索时习惯使用数据库系统默认的检索项。以 ScienceDirect 数据库为例，系统默认的检索项是在全文中进行检索（All Fields）（图 12.1）。这些默认的检索项并不能满足所有的检索需求。合理地选择检索项，可以使检索结果更为科学。例如，当检索结果过多时，可以考虑使用范围较窄的检索项，如题目（title）、关键词（keywords）、摘要（abstract）等，从而精简检索结果；反之，则可以考虑使用范围较宽的检索，从而扩大检索结果。文献年份默认的是从 2007 年到现在（一般提供的是近十年的文献），那么，如果想了解最

近几年某一方面的研究进展，则可以在选择列表中进行选择。

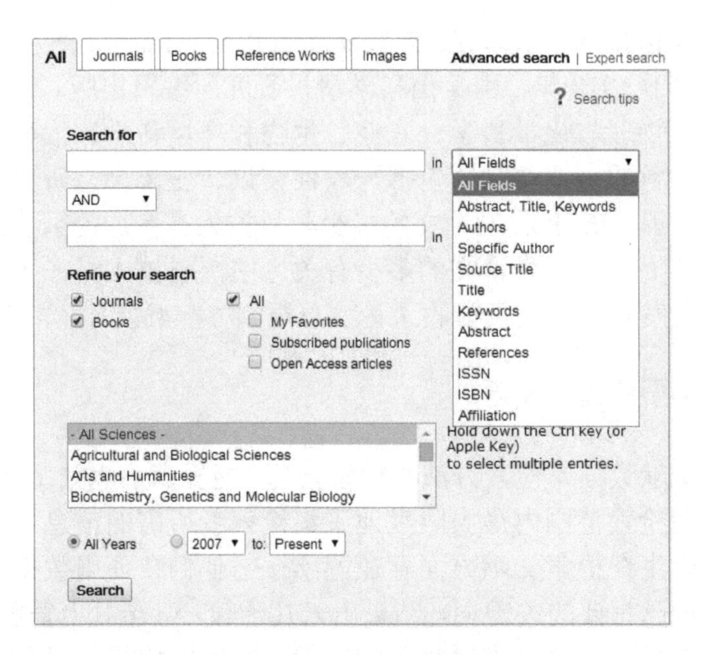

图 12.1　ScienceDirect 数据库检索首页

2. 查不到全文

国内好多高校没有购买外文全文数据库，这导致学生在查英文文献时存在一定困难，下面介绍几种获取英文文献全文的方法。

（1）可以利用一些免费文献网站（如 Highwire press、wikipedia、literature 等）和一些开放杂志主页。

（2）可以通过 e-mail 直接向作者索取。

（3）可以通过一些中转网站进行索取，如通过目前国内使用最多的“百链”进行搜索。

Unit 13 Composition and writing of English scientific papers

Lesson 51 Composition of English scientific papers

科技论文在情报学中又称为原始论文或一次文献，它是科学技术人员或其他研究人员在科学实验（或试验）的基础上，对自然科学、工程技术科学，以及人文艺术研究领域的现象（或问题）进行科学分析、综合的研究和阐述，进一步进行一些现象和问题的研究，总结和创新另外一些结果和结论，并按照各个科技期刊的要求进行电子和书面的表达。因此，所谓的英语科技论文即是用英语写的科技论文。

英语科技论文的组成

科技论文的分类就像它的定义一样，有很多种不同的分法。科技论文可从不同的角度，根据不同标准进行分类。本部分主要以一般研究性论文为例讲解英语科技论文的组成。

英语科技论文的基本组成包括：

题目（title）

作者（author）

联系方式（affiliation（s）and address（es））

摘要（abstract）

关键词（keywords）

前言（introduction）

材料与方法（materials and methods）

结果（results）

讨论（discussion）
结论（conclusion）
致谢（acknowledgements）
参考文献（references）
附录（appendices）等。

下面将以一篇具体文章为例讲解每一部分的写作要求，请参考本书末附录Ⅱ所附范文。

Lesson 52 Title, author and contact information of the paper

一、题目（title）

科技论文的标题是其基本思想的浓缩与概括。一个好的标题应该确切、鲜明、扼要地概括论文的基本思想，使读者在未看论文的摘要和正文之前即能迅速准确地判明论文的基本内容，从而做出是否阅读摘要和正文的判断。那么，如何提炼一个既符合国际标准又能概括论文基本思想的标题呢？具体要求如下：

（1）标题的主要目的是吸引读者和协助检索，因此一定要包含一些关键词。

（2）论文题目一般由名词词组或名词短语构成，一般不用句子表示。在必须使用动词的情况下，一般用分词或动名词形式。

（3）标题中的大小写，标题字母中的大小写根据不同期刊的要求一般有以下 3 种格式：①全部字母大写。如 CHANGES IN PHOTOSYNTHESIS AND ANTIOXIDANT DEFENSES OF PICEA ASPERATA SEEDLINGS TO ENHANCED ULTRAVIOLET-B AND TO NITROGEN SUPPLY。②每个词的首字母大写，但冠词、连词和介词全部小写。如 Changes in Photosynthesis and Antioxidant Defenses of Picea Asperata Seedlings to Enhanced Ultraviolet-B and to Nitrogen Supply。③标题第一个词的首字母大写，其余字母均小写（拉丁名首

字母除外）。

（4）标题中一般不适用缩略词、化学分子式、专利商标名称、行话、罕见的或过时的术语。

（5）标题中一般不使用一些没有实际意义的词，如 Some、a few、a study of 等。

二、作者（author）

文章作者一栏位于标题行之后，国家标准 GB7713-87《科学技术报告、学位论文和学术论文的编写格式》规定，在学术论文中署名的作者“只限于那些选定研究课题和制定研究方案、直接参加全部或主要部分研究工作并作出主要贡献，以及参加撰写论文并能对内容负责的人，按其贡献大小排列名次”。

国际医学期刊编辑委员会（ICMJE）有关作者资格的界定包括三条：第一，参与课题的构思与设计，资料的分析和解释；第二，参与论文的撰写或对其中重要学术内容作重大修改；第三，参与最后定稿，并同意投稿和出版。必须同时具备这三个条件才能成为作者；作者的排列顺序应按贡献大小来排列，每位作者都必须就论文的全部内容向公众负责。

1. 中国人名格式

按照欧美国家的习惯，名字（first name）在前，姓氏（surname / family name / last name）在后。通过查阅发现，当前，中国人在国际期刊上发表文章时人名的表达方式主要有以下几种格式（以张三丰为例）：Sanfeng Zhang、San-Feng Zhang、Sanfeng ZHANG、San-Feng ZHANG、Zhang Sanfeng、Zhang San-feng，其中最常见的写作方式是 Sanfeng Zhang 和 San-Feng Zhang。一般国外期刊会尊重作者对自己姓名的表达方式，但前提一定要告诉编辑部，否则编辑部会默认为放在后面的是姓。不管作者采用哪种写作方式，全文应该保持一致。

2. 署名注意事项

（1）论文的执笔人或主要撰写者应该是第一作者，例如，一般来讲研究生写的论文应以研究生为第一作者，而不能以导师为第一作者。

（2）不可故意将知名人士署为作者之一，要避免“搭车”现象。

（3）不能故意遗漏具有署名权利的作者。

（4）坚持原则：既不要随便增加作者，也不要随便被增加。

三、联系方式 (affiliation(s) and address(es))

在作者姓名的下方还应注明作者的工作单位、邮政编码、电子邮件地址或联系电话等。也有刊物在论文标题页的页脚标出以上细节，在论文最后附上作者简介和照片。该信息要求准确清楚，其目的一方面使读者能按所列信息顺利地与作者联系；另一方面，有利于不同单位每年成果的统计和考核。

主要事项：

（1）单位一般都写到二级单位，要参照单位对外正式名称，一般不用缩写。

例如：

The College of Life Sciences, Hebei University, Baoding 071002, China;

Chengdu Institute of Biology, Chinese Academy of Sciences, Post Box 416, Chengdu 610041, China.

（2）如果由多个单位共同完成，要根据贡献大小排列顺序。

Lesson 53 The writing skills of abstract and keywords

一、摘要（abstract）

摘要也称内容提要，是对论文的内容不加注释和评论的简短陈述。一般位于正文之前，其作用主要是为读者阅读、信息检索提供方便。摘

要不宜太详尽，也不宜太简短，应将论文的研究体系、主要方法、重要发现、主要结论等简明扼要地加以概括。摘要对于写作新手而言比较难，尤其是对于大学生和研究生，由于他们大学期间主要以写实验报告为主，因此，在写摘要时总是摆脱不了实验报告模式的束缚，其实如果明白摘要的主要构成要素，摘要写作还是比较容易掌握。

1. 部分摘要的构成要素

摘要由四部分构成：研究目的、研究方法、研究结果和结论。

（1）研究目的——准确描述该研究的目的，说明提出问题的缘由，表明研究的范围和重要性。

（2）研究方法——简要说明研究课题的基本设计，结论是如何得到的。

（3）研究结果——简要列出该研究的主要结果，有什么新发现，说明其价值和局限。叙述要具体、准确并给出结果的置信值。

（4）结论——简要地说明经验，论证取得的正确观点及理论价值或应用价值，是否还有与此有关的其他问题有待进一步研究，是否可推广应用等。

范例：

The paper mainly studied the effects of ultraviolet-B (UV-B) radiation, nitrogen, and their combination on photosynthesis and antioxidant defenses of *Picea asperata* seedlings（研究目的）. The experimental design included two levels of UV-B treatments (ambientUV-B,11.02 KJ m^{-2} day^{-1} ; enhanced UV-B,14.33 KJ m^{-2} day^{-1}) and two nitrogen levels (0; 20 $gm^{-2}a^{-1}N$) to determine whether the adverse effects of UV-B are eased by supplemental nitrogen（研究方法）. Enhanced UV-B significantly inhibited plant growth, net photosynthetic rate (A), stomatal conductance to water vapor (Gs), transpiration rate and photosynthetic pigment, and increased intercellular CO_2 concentration, UV-B absorbing compounds,proline content, malondialdehyde (MDA) content, and activity of antioxidant enzymes (peroxidase (POD), superoxide dimutase, and glutathione reductase). Enhanced UV-B also reduced needle DW and increased hydrogen peroxide (H_2O_2) content and the rate of superoxide radical (O_2^-) production only under supplemental nitrogen. Supplemental nitrogen increased plant growth, A, Gs, chlorophyll content and activity of antioxidant enzymes (POD, ascorbate peroxidase, and catalase), and reduced MDA content, H_2O_2 content, and the rate of O_2^- production only under ambient UV-B, whereas supplemental nitrogen reduced activity of antioxidant enzymes under enhanced UV-B. Carotenoids content, proline content, and UV-B absorbing compounds increased under supplemental nitrogen. Moreover, significant UV-B×nitrogen interaction was found on plant height, basal diameter, A, chlorophyll a, activity of antioxidant enzymes, H_2O_2, MDA, and proline content（结果）. These results implied that supplemental nitrogen was favorable for photosynthesis and antioxidant defenses of *P. asperata* seedlings under ambient UV-B. However, supplemental nitrogen made the plants more sensitive to enhanced UV-B, although some antioxidant indexes increased（结论）.

2. 摘要撰写要求

（1）摘要应具有独立性和自明性，并拥有一次文献同等量的主要信息，即不阅读文献的全文，就能获得必要的信息。因此，摘要是一种可以被引用的完整短文。

（2）英文摘要的时态：摘要所采用的时态因情况而定，应力求表达自然、妥当。写作中可大致遵循以下原则：介绍背景资料时，如果句子的内容是不受时间影响的普遍事实，应使用现在时（范例中的下划线句子）；如果句子的内容为对某种研究趋势的概述，则使用现在完成时（如 Recent researchers have identified that selenium (Se) could not only promote growth and development of the plant, but also increase resistance and antioxidant capacity of the plant subjected to stress, although Se is not considered to be required by higher plants）；叙述作者的工作和结果一般用过去式（见范例）。

范例：

Enhanced ultraviolet-B (UV-B) is one of the most important abiotic stress factors that can influence almost every aspect of plant. Selenium (Se) can increase the tolerance of plants to stressful environment. The paper mainly reported the effects of enhanced UV-B, Se supply and their combination on agronomical characters of winter wheat under field conditions. Enhanced UV-B caused a marked decrease in chlorophyll

content, plant height, spike length, weight per spike, grain yield and protein content, grain nitrogen (N) and iron (Fe) concentration, and increased hydrogen peroxide (H_2O_2), malondialdehyde (MDA) and proline content, and grain zinc (Zn) and manganese (Mn) concentration under without supplemental Se supply. However, it also decreased plant height, spike length, weight per spike, grain yield and Fe concentration, and increased H_2O_2 content, grain potassium (K), Zn and Mn concentration under supplemental Se supply. Se supply induced an evident increase in chlorophyll content, spike length, weight per spike, grain yield, grain protein content, grain N, Fe, copper (Cu), and Se concentration under both UV-B levels. Moreover, significant UV-B×Se interaction was found on plant height, chlorophyll, MDA, H_2O_2 and proline content, and grain protein, N, K, Cu and Mn concentrations in wheat. The obtained results supported the hypothesis that Se supply increased the yield and improved the quality of winter wheat exposed to enhanced UV-B to some extent （Yao XQ, Chu JZ, He XL, et al. 2013）。

（3）英文摘要不宜超过 250 个实词，要求结构严谨、语义确切、表述简明、一般不分段落。

（4）不使用非本专业的读者尚难以清楚理解的缩略语、简称、代号，如确有需要（如避免多次重复较长的术语）使用非同行熟知的缩写，应在缩写符号第一次出现时给出其全称；不使用一次文献中列出的章节号、图号、表号、公式号及参考文献号。

（5）应尽量避免使用图、表、化学结构式、数学表达式、角标和希腊文等特殊符号。

3. 英文摘要常见的主题句

（1）起始句：

This paper / article / mainly studied/introduced/discussed/analyzed

（2）结尾句：

The results indicated /showed/suggested that …

The obtained results supported the hypothesis that…

二、关键词（keywords）

关键词属于主题词的一种，是作者在完成论文写作后，选出的能表示论文主要内容的词汇。关键词既可以作为文献检索或分类的标识，它本身又是论文主题的浓缩。读者从中可以判断论文的主题、研究方向和方法等。在提炼关键词时应注意以下几点：

（1）关键词一般是名词或名词词组，可以从论文标题和摘要中选择。

（2）不要使用过于宽泛的词做关键词（如 research、methods、effects、study 等）。

（3）避免使用自定的缩略语、缩写词作为关键词，除非是科学界公认的专有缩写词（如 DNA）。

（4）关键词的数量要适中，一般以 3～8 个词作为关键词。不同期刊的要求不同，关键词一般以分号、逗号或空格分开。

（5）关键词的首字母一般大写，并且按关键词的首字母顺序排列。

Lesson 54　The writing skills of the introduction

前言 (introduction) 位于正文的起始部分，是论文中最重要的组成部分之一，科技论文中前言的根本目的是向读者和审稿专家阐明“为什么要做这个研究”。英语科技论文中的前言内容较长，通常被认为是论文中比较难写的部分之一。那么，如何把前言写好呢？下面就以本书末附录Ⅱ所附范文中的前言为例来讲具体写作过程。

在讲前言写作前，首先交代一下范文的主要研究内容，以便于读者更好地理解前言的写作过程。本书中所使用的范文主要研究了 UV-B 辐射增强和氮增加对云杉幼苗的光合作用和抗氧化防御的影响。这个主要内容可能会首先让人想到如下几个问题：作者为什么要做这个研究？这个研究与相关研究有什么不同（即创新点），本研究主要做了哪些工作？做这个研究有什么意义？下面带着这几个问题再来探讨如何写前言。

1. 前言包括的主要内容

（1）作者为什么要做这个研究？进一步解释为什么要研究 UV-B 辐射和氮供应？那么，要让读者明白这个问题，首先就得阐述一下自己研究领域的基本内容，也就是背景资料的介绍，参考范文的 Introduction 中第 1 和第 2 段分别介绍了 UV-B 辐射和氮沉降的基本情况。

（2）问题的提出，也就是作者做这个研究的创新之处。请参考范文的 Introduction 中的第 3 段。第 3 段主要讲了 UV-B 辐射和氮沉降在植物上的研究现状，通过对研究现状的概括总结提出先前研究中存在的不足或者尚未解决的问题。

（3）解决第 2 个问题中提出的问题，即研究内容和假设，请参考范文的第 4 段。

（4）所谓的研究意义，就是针对所做研究做出一个简单客观的评价。请参考范文中的第 4 段。

2. 前言写作要求

（1）尽量准确、清楚且简洁地阐述自己研究领域的基本内容。

（2）文献要尽量全面客观：不要对相似文献避而不引，或者只引别人早期的工作，这会让人感到你对该领域了解不够，甚至是有意回避，这不符合学术规范。不能过分贬低别人的研究工作。

（3）给自己的研究工作一个恰如其分的地位。既不要夸大也不要缩小。

（4）一般不进行展开讨论。

（5）参考文献必须是与本篇论文相关的文献，最好有近三年的文献。

范例：

Atmospheric ozone remains depleted and the annual average ozone loss is approximately 3% globally (Executive summary 2003). Researches have shown that enhanced ultraviolet-B (UV-B) reaching the surface of the earth has very many adverse impacts on plant growth (Jordan 1996, 2002, Jansen 2002). When plants are exposed to UV-B stress, they could induce some protective mechanisms. For example, the increases in UV-B absorbing compounds, proline content, and activity of antioxidant enzymes have been reported (Baumbusch et al. 1998, Prochazkova et al. 2001, Saradhi et al. 1995).

In addition to UV-B radiation, human activities have significantly altered the global nitrogen cycle, with the development of industry and agriculture. More and more nitrogen will be imported into the terrestrial ecosystems through nitrogen deposition. In the European livestock and industrialized areas, nitrogen deposition was more than 25 kg hm^{-2} a^{-1} N (Binkley et al. 2000). In the Northeastern United States, the current nitrogen deposition was more 10–20 times than nitrogen in background (Magill et al. 1997). At present, China has been one of three high-nitrogen deposition regions (Li et al. 2003).

Nitrogen is the mineral nutrient needed in largest amounts by plants and it is usually also the limiting factor for plant growth in terrestrial ecosystems (Vitousek and Howarth 1991), particularly in tundra, boreal as well as alpine ecosystems (Xu et al. 2003). At the same time, nitrogen is also an important constituent of photosynthetic apparatus (Correia et al. 2005). Maximum photosynthetic capacity is strongly regulated by leaf nitrogen concentration (Field and Mooney 1986). In contrast to UV-B radiation, supplemental nitrogen improved growth and net photosynthesis of plant (Nakaji et al. 2001, Keski-Saari and Julkunen-Tiitto 2003) and reduce production of free radicals in plants (Ramalho et al. 1998). UV-B radiation and nitrogen are expected to increase simultaneously with future changes in global climate. Nitrogen can affect UV-B response in plants (Correia et al. 2005, Pinto et al. 1999). Previous studies have mainly focused on crop and herb plants, although forests account for over two-thirds of global net primary productivity (NPP), compared with about 11% for agricultural land (Barnes et al. 1998). However, only limited papers have been reported on the combined effects of nitrogen and UV-B radiation on woody plants (De La Rose, et al. 2001, 2003; Lavola, et al. 2003).

Picea asperata is a key species in the southeast of the Qinghai-Tibetan Plateau of China and widely used in reforestation programs at present (Liu 2002). The paper mainly studies the short-term

influence of enhanced UV-B radiation and supplemental nitrogen on photosynthesis and antioxidant defenses of *P. asperata* seedlings under semi-field condition. This will be helpful for understanding of the combined effects on conifer tree species and development of improved plant tolerance toward stressful environmental factors. On the basis of previous study in other species, we hypothesized that (1) both UV-B and nitrogen would affect photosynthesis and antioxidant defenses of *P. asperata* seedlings; and (2) supplemental nitrogen modifies the adverse effects of UV-B on the conifer plants, in order to better understand the responses of woody plant to both enhanced UV-B and to supplemental nitrogen in future.

Lesson 55 The writing skills of materials and methods

对于科学论文而言，这部分主要用来介绍试验对象、条件、使用的材料、试验设计、参数测定或计算的过程、公式的推导、模型的建立及数据统计方法等。这部分要求提供足够的细节，使别人能够按照该方法重复这些试验，如果采用的是标准方法，要有参考文献，并且给出关键操作步骤。这部分在论文中的位置通常位于 introduction 后面，但有部分期刊要求放在“讨论和结论”后面。通常在英语科技论文中见到的格式如下：

范例 1：materials and methods
plant material and experiment design
UV-B treatments and nitrogen treatments
pigment analysis
…
statistical analysis

范例 2：materials and methods
study site
experiment design

sampling and sample analysis
data analyses

主要内容

1. 试验材料

这部分主要介绍试验条件和试验材料，试验材料要作详细的说明，应写明材料的来源、数量和规格。对于试验用的动物、植物和微生物要正确写明它们的名称，还要写明它们的特征（年龄、性别、遗传学和生理学上的状况）。如本课末所附范文中 Materials and methods 部分中的 Plant material and experiment design 中的第 1 段：

The experiment was conducted in open semi-field condition from April 15 to October 15, 2005 in Maoxian Ecological Station of Chinese Academy of Sciences, Sichuan province, China (31°41′ N, 103°53′ E, 1820m a.s.l.). Four-year-old *P. asperata* seedlings were from a local nursery. The plant height, basal diameter and whole-plant FW at the beginning of the experiment were 15.38±0.48cm, 6.52±0.35mm, and 7.52±0.43g, respectively. Seedlings were transplanted into plastic pots (25cm diameter and 35cm depth) with a 12h photoperiod and a daily average 1200 mmol m^{-2} s^{-1} photosynthetic photon flux density (PPFD), one seedling per pot. The substrate used for growing the seedlings was sieved topsoil from a spruce-forest. In a preliminary experiment, the plastic pots did not affect growth of seedling root during a 2-year growth period.

2. 试验设计

这部分要详细介绍自己的试验设计过程，即该试验是如何做的，必须给出细节，让别人能够按照该试验描述进行重复。如果试验设计中用到一些浓度或数量，要给出使用该浓度或数据的依据。请参考本书末附录 II 所附范文中 Materials and methods 中的 Plant material and experimental design 中的第 2 段和 UV-B treatments and nitrogen treatments 部分。

Experiment design

The experiment consisted of four treatments in the paper: (1) ambient UV-B without supplemental nitrogen (control, C); (2) ambient UV-B with supplemental nitrogen (N); (3) enhanced UV-B without supplemental nitrogen (UV-B); and (4) enhanced UV-B with supplemental nitrogen (UV-B+N). Each treatment has three blocks and each block has 10 pots. The pots within blocks were rotated approximately every 20 days.

UV-B treatments and nitrogen treatments

Supplementary UV-B was supplied by UV-B fluorescent lamps (Beijing Electronic Resource Institute, Beijing, China) mounted in metal frames with minimum shading. The distance from the lamps to the top of plant apex was 100 cm and kept constant throughout the experiment. In ambient UV-B frames, UV-B from the lamps was excluded by wrapping the tubes with 0.125mm polyester film (Chenguang Research Institute of Chemical Industry, Chengdu, China), which transmits UV-A. In enhanced UV-B frames, lamps were wrapped with 0.10mm cellulose diacetate film, which transmits both UV-B and UV-A. Vertical polyester curtains were placed between the frames in order to prevent the UV-B radiation from reaching the C seedlings (De La Rose et al. 2003). Films were replaced every week. The lamp duration was modified monthly and replaced in times. The spectral irradiance from the lamps was determined with an Optronics Model 742 (Optronics Laboratory Inc., Orlando, FL) spectroradiometer. The spectral irradiance was weighted according to the generalized plant action spectrum (Caldwell 1971) and normalized at 300 nm to obtain effective radiation ($UV\text{-}B_{BE}$). The supplemental UV-B BE dose was 3.31 KJ m^{-2} day^{-1} (a 30% difference in ambient $UV\text{-}B_{BE}$) in addition to the effective 11.02 KJ m^{-2} day^{-1} UV-B BE (ambient UV-B BE) from sky. All pots also received natural solar radiation. Seedlings were irradiated for 8 h daily centered on the solar noon.

Nitrogen was added as 9.5 m*M* NH_4NO_3 solution (300 ml) to the potted soil surface every 3 days. The treatment without supplemental nitrogen was watered with 300 ml of water. The nitrogen amount added to the soil was equivalent to 20 g m^{-2} a^{-1} N on the basis of soil surface area. Nitrogen amount was based on the similar studies (Bowden et al. 2004; Nakaji et al. 2001).

3. 参数测定方法

对每一种指标的测定方法描述要详略得当、重点突出，所采用的方法必须是公开报道过的方法，需要写明引用相关的文献或者实验指导书，如果在试验过程中对先前已公开出版的方法进行了少量的修改，写作过程中要注明。请参考本书末附录II所附范文中 Materials and methods 中参数的测定。

Pigment analysis

Samples of the youngest, fully expanded needles were taken for the determination of chlorophyll content. Needle was grinded in 80% acetone for the determination of chlorophyll and carotenoids (Car). Total chlorophyll [Chl (*a*+*b*)], chlorophyll a (Chl *a*), chlorophyll *b* (Chl *b*) and total Car contents were determined according to Lichtenthaler (1987).

The rate of superoxide radical production and hydrogen peroxide content

The rate of superoxide radical production (O_2^-) was measured as described by Ke et al. (2002), by monitoring the nitrite formation from hydroxylamine in the presence of O_2^-. Needles (0.5g) were homogenized with 1.5 ml of 65-m*M* potassium phosphate (pH 7.8) and centrifuged at 5000 g for 10 min. The incubation mixture contained 0.45 ml of 65-m*M* phosphate buffer (pH 7.8), 0.5 ml of 10 m*M* hydroxylamine hydrochloride, and 0.5 ml of the supernatant. After incubation at 25°C for 20 min, 8.5 m*M* sulfanilamide and 3.5 m*M* α-naphthylamine were added to the

incubation mixture. After reaction at 25°C for 20 min, the absorbance in the aqueous solution was read at 530 nm. A standard curve with NO_2^- was used to calculate the production rate of O_2^- from the chemical reaction of O_2^- and hydroxylamine.

4. 数据统计

该部分主要介绍论文中所有数据处理采用何种软件，使用何种处理方法。

范例 1：

Statistical analysis

All data were subjected to an analysis of variance that tested the UV-B radiation, nitrogen and UV-B×nitrogen interaction effects, and the significance of the single factors calculated as well as the interaction between the factors calculated. All statistical analyses were performed using the Software Statistical Package for the Social Science (SPSS) version 11.0 (SPSS Inc., Chicago, IL).

范例 2：

Statistical analysis

All analyses were performed using SPSS 17.0. Before analysis, all data were tested for the homoscedasticity. If data were heterogeneous, they were ln-transformed before analysis. A two-way analysis of variance was used to test the effects of warming, forest type and their interactions on all of the variables. For specific forest type, Student t-tests were used to compare the effect of the experimental warming. We also used Pearson's correlation analyses to examine the relationships between ECM colonization, ECM fungal biomass, plant tissue N concentrations, plant tissue biomass, root physiology parameters (RV and NR activity) and rhizosphere soil inorganic N (NH_4^+-N and NO_3^--N). The statistical tests were considered significant at the $P < 0.05$ level (Li YJ, Sun DD, Li DD, et al. 2015).

Lesson 56 The writing skills of results, discussions and conclusions

一、结果（results）

结果部分是论文的核心，是作者通过试验观测或调查所得出的结果及各种图像和数据资料。这部分在写作过程中要简明扼要，合理展示数据，并能证明研究结果与目的一致或相反，假设是否成立。合理运用统计，误差分析；检查数据的准确性和一致性。这部分在论文中多数情况是单独作为一节来写，有部分期刊要求“Results and Discussions”放在一起来写，即边写边讨论。下面以“Results”单独作为一节为例介绍这部分写作需要注意的事项。

（1）结果中图表要求。一篇论文通常会包含很多数据，这些数据需要用图或表表示出来，国际期刊要求文中的表一般用三线表。请参考范文中的 Tables 和 Figures。很多读者在查阅文献时不明白图和表中数据所包含的意思，下面将以本书末附录II所附范文中的 Table 3 为例简要介绍。

圆圈中的数据为“0.35 ± 0.01^{b}”，这是通过 SPSS 分析的结果，其中“0.35”表示几个重复的平均值，“±0.01”表示的是几个重复值之间的正负误差，“b”表示的是不同处理之间的显著性。例如，我们来看表中第三列中的数据（方格中的数据），数据后面的字母表示的是不同处理之间的显著性差异，如果包含相同字母则表示处理间没有显著差异，包含不同字母则表示处理间具有显著差异。

Table 3. The effects of enhanced UV-B and supplemental nitrogen on photosynthetic pigment of *P. asperata*. Values are the mean ± SE of six replicates in column rows 1–4 and the values in the same column with different letters are significantly different from each other ($P < 0.05$). Significant effects of the two factors as well as of the interaction are indicated in column rows 5–7.

Treatment	Chl *a* (mg g^{-1} FW)	Chl *b* (mg g^{-1} FW)	Chl *a*/*b*	Chl (*a* + *b*) (mg g^{-1} FW)	Car (mg g^{-1} FW)
C	0.35 ± 0.01^{b}	0.10 ± 0.00^{b}	3.50 ± 0.38^{a}	0.45 ± 0.02^{b}	0.08 ± 0.00^{b}
N	0.43 ± 0.00^{a}	0.12 ± 0.00^{a}	3.58 ± 0.31^{a}	0.55 ± 0.00^{a}	0.10 ± 0.00^{a}
UV-B	0.28 ± 0.01^{c}	0.07 ± 0.01^{c}	4.00 ± 0.49^{a}	0.35 ± 0.05^{c}	0.04 ± 0.00^{d}
UV-B + N	0.30 ± 0.00^{c}	0.08 ± 0.01^{c}	3.75 ± 0.24^{a}	0.37 ± 0.00^{c}	0.06 ± 0.00^{c}
N	0.000	0.041	0.132	0.040	0.001
UV-B	0.000	0.000	0.085	0.000	0.000
N × UV-B	0.001	0.405	0.320	0.124	0.531

（2）要实事求是地汇报结果或数据无须加入自己的解释，即使得到的结果与试验不符，也不可不写，而且还应在讨论中加以说明和解释。要根据图或表中统计分析后的数据来写。请参考本书末所附范文中和 Table 3 中的数据相对应的文字部分。注意处理间有没有显著差异是根据统计结果来分析的，而不是根据数据的大小。

范例：

Enhanced UV-B markedly reduced Chl *a*, Chl *b*, Chl (*a*+*b*), and Car content (Table 3). On the other hand, Chl *a*, Chl *b*, and Chl (*a*+*b*) content of plants grown at ambient UV-B were increased by supplemental nitrogen, whereas supplemental nitrogen did not influence chlorophyll pigment under enhanced UV-B. Car content was increased by supplemental nitrogen. A parallel change trend in Chl *a* and Chl *b* resulted in no significant change in Chl *a*/*b* ratio under enhanced UV-B or supplemental nitrogen. Significant interactive effects of UV-B and nitrogen were also detected on Chl *a* content ($P < 0.001$).

（3）文字表达应准确、简洁、清楚，应在句子中指出图表所揭示的结论，并把图表的序号放入括号中，让读者清楚地知道所描述的结果是哪个图。例如，Enhanced UV-B markedly reduced Chl *a*, Chl *b*, Chl (*a*+*b*), and Car content (Table 3).

（4）这部分主要是叙述或总结研究结果，常用一般过去时。

二、讨论（discussion）

讨论是论文的升华部分，也是论文中比较难写的部分之一，讨论的一个重要作用就是要突出自己研究的创新性，体现出显著区别于他人的特点。要突出自己研究的创新性，就需要和相关的研究进行比较，因此，这部分需要引用大量的参考文献。请参考本书末附录II所附范文。

1. 讨论部分主要内容

（1）第一段主要概述自己的研究目的和假设。选择要深入讨论分析的问题，对于和前人研究相一致的结果，就不需要再深入讨论，而对于自己研究中新的发现或与相关研究不同的结果要进行重点讨论和分析原因。

（2）指出根据自己的研究结果所得出的结论或推论。

（3）指出本研究所受到的限制及这些限制对研究结果可能产生的影响。并对进一步的研究题目或方向提出建议。

（4）指出自己研究的理论意义和实际应用价值。

2. 讨论写作过程中需注意事项

（1）对结果的解释要重点突出、简洁、清楚，结论需严格、客观。

（2）在 discussion 中，需要特别指出的是要保持和 results 的一致性，也就是讨论和结果要一一对应，前后呼应，相互衬托。切勿出现按照讨论的内容推出与试验相反的结论这种事情。

（3）discussion 部分只对本文结果进行讨论，与同类研究结果进行比较。

（4）对结果科学意义和实际应用效果的表达要实事求是，不能过分夸大，也不能太谦虚。

（5）时态的运用：回顾研究目的和概述研究结果时，通常使用过去时，阐述由结果得出的推论时，若为普遍有效的结论或推论（而不只是在讨论自己的研究结果），并且结果与结论或推论之间的逻辑关系不受时间影响，通常使用现在时。

三、结论（conclusion）

结论是作者对研究的主要发现和成果进行概括总结，让读者对全文的重点有一个深刻的印象。结论应完整、准确、鲜明地表达作者的观点。它同引言相呼应，而且和摘要一样，可以使读者不通读全文即可掌握论文的要旨。在 conclusion 部分，作者应该总结阐明论文的主要结果及其重要性，有的文章也在本部分提出当前研究的不足之处，对研究的前景和后续工作进行展望。这部分在论文写作中有时单独列为一节，有时放在“讨论”或“结果与讨论”中。

结论写作中需注意事项

（1）结论要来源于论文，不能编造无法从论文中导出的结论。

（2）要与前言相呼应，不能模棱两可。

（3）不做自我评价。

（4）结论中总结结果语句一般用过去时，对于展望语句用将来时。

（5）常见的结尾句：It can be concluded that…；The following conclusions can be drawn from…；In conclusions…

范例 1：

UV-B-effects on biochemical traits in postharvest flowers depended on UV-B radiation levels. The results indicated that optimal UV-B radiation could promote secondary metabolism processes and increase medically active ingredients in postharvest flowers. Proteomic analysis revealed that 19 differentially expressed protein spots were successfully indentified by MALDI-TOF MS. These proteins were mainly involved in photosynthesis, respiration, protein biosynthesis and defence, and secondary metabolism. This study provides new insights into the responses in postharvest flowers to enhanced UV-B radiation. Further studies are needed to better understand the molecular basis of the UV-B effects on medically active ingredients in postharvest plant organs.(See: Yao et al. 2015).

范例 2：

In conclusion, enhanced UV-B led to a marked decline in wheat yield and protein concentration, and influenced nutritional element concentration of wheat grain. Se supply increased wheat yield and protein concentration, and increased most micro-element concentration in wheat grains, which further improved quality of wheat subjected to UV-B stress to some extent. Our results confirmed our hypothesis (See: Yao et al. 2013).

范例 3：

In conclusion, the present study demonstrated that four-year experimental warming decreased ECM colonization and biomass, root vigor, and N concentration of most plant components, but increased the biomass and N concentration of the uptake organ (fine root) in natural forest, and consequently total N content of P. *asperata* seedlings were significantly increased. However, ECM colonization and plant N accumulation of the seedlings in plantation were insensitive to four-year OTC-warming. The different responses to warming in the two contrast forests would bring two disparate growth potential to the seedlings. In addition, the changes of ECM colonization and fine root biomass for effective N uptake was good for transferring soil N to plant N pool, and potentially remit the N leaching under future warming in natural forest ecosystem.(Li YJ, Sun DD, Li DD, et al. 2015).

Lesson 57 The writing skills of acknowledgements and references

一、致谢（acknowledgements）

科学研究工作常常需要多方面的指导和帮助才能完成，因此，当科研成果以论文形式发表时，有时需要对他人的劳动给予充分肯定，并郑重地以书面形式表示感谢。对于致谢的对象，可直书其名，也可写尊称，如某某教授、某某博士等。其顺序最好依贡献的大小来排列，而不要依年龄、地位排列。致谢的言辞应该恳切、实事求是、恰如其分，而不应浮夸或单纯地客套。致谢的语句要尽量简短。

1. 致谢对象和范围

（1）在试验过程中给予技术支持的单位、团体或个人。

（2）给予经费资助的单位、团体或个人（如国家自然科学基金、省自然科学基金等）。

（3）在论文选题、构思、试验设计、数据收集及论文撰写过程中提供建设性的意见及提供指导的老师、同事或同学。

（4）提供过试验材料、仪器，及给予其他方便的人。

2. 致谢的写作注意事项

（1）致谢的内容应尽量具体，一般置于结论之后，参考文献之前。

（2）致谢的文字表达要朴素、简洁，以显示其严肃和诚意。

范例 1：

Acknowledgements — During this work, the senior author was supported by the National Natural Science Foundation of China (No. 30530630), the Talent Plan of the Chinese Academy of Sciences and Knowledge Innovation Engineering of the Chinese Academy of Sciences (Yao XQ, Liu Q. 2007).

范例 1 中作者主要致谢了资金资助单位。

范例 2：

Acknowledgements—We are very grateful to G. Glauser (University of Neuchâtel, Switzerland) for the analysis of prenylquinones. We also thank V. Burlat and V. Courdavault for providing the BiFC vectors, M. Rothbart and B. Grimm for the GGR clone, P. Obrdlik for the Y2H plasmids, F. Wüst for his support with the Y2H setup and M. Nater for her phenotyping work. Technical support from M.R. Rodríguez- Goberna and members of the CRAG Services is greatly appreciated. We would finally like to thank M. Meret from the Max Planck Institute of Molecular Plant Physiology in Golm (Germany) for support in the metabolomics sample preparation procedure. This work was supported by grants from CYTED (Ibercarot-112RT0445), MINECO (BIO2011-23680), AGAUR (2014SGR-1434), ETH Zurich (TH-51 06-1), the EU FP7 contract 245143 (TiMet) and SK grant VEGA 1/0417/13. M.A.R-S. and M.V.B. were supported by MINECO FPI and AGAUR FI fellowships (Ruiz-Sola et al. 2016).

范例 2 中作者主要感谢了帮他们测定参数的一些人和论文评审者，同时也感谢了项目资金提供者。

二、参考文献（references）

参考文献是科技论文的重要组成部分，凡在论文中引用前人已发表的文献中的观点、数据和材料等均应对它们在正文中出现的地方予以标明并在文末列出参考文献。参考文献的主要作用是反映论文的科学依据，表现作者对他人研究成果的尊重，向读者提供文中引用有关资料的出处，或为了节约篇幅或叙述方便，提供在论文中涉及而没有展开的有关内容的详尽文本。

参考文献的内容可包括书籍、期刊文章、学位论文、会议资料、电子文献资源等，一般都要包含作者姓名、作品名称和出版信息等内容。

1. 参考文献格式

参考文献格式一定要严格按照所投期刊的“Submission guidelines”或“guidelines for authors”中对参考文献的要求。不同期刊对参考文献的格式要求略有不同，投稿者可参考同一刊物最新刊出论文中参考文献的格式，使自己论文的文献列举和标注方法与所投刊物相一致。

文中参考文献的标注有两种（此处所举示例均依据所投期刊格式要求）：

（1）根据参考文献在文中出现的先后顺序按阿拉伯数字序号编排。例如：Researches have shown that enhanced ultraviolet-B (UV-B) reaching the surface of the earth has very many adverse impacts on plant growth[1, 2]。此种表示法需要注意三点：①有的期刊要求用上标表示；②如果一句话有 3 个或 3 个以上参考文献，通常用如下表示方法：（5-8）；③每一篇引用的参考文献都有一个固定的序号，如果同一篇参考文献在文中多次引用，则都用该文献在文中第一次出现时的序号。

（2）用所引用文献中作者姓和文献发表年代来表示。例如：Researches have shown that enhanced ultraviolet-B (UV-B) reaching the surface of the earth has many adverse impacts on plant growth (Jordan, 1996, 2002; Jansen, 2002)。这种表示方法需要注意两点：①如果所引用参考文献只有一位作者，一般表示方法为作者姓加发表年代；如果有两位作者，一般用两位作者姓加发表年代，如 Jordan and Jansen, 2015；如果有三位及以上作者，一般用第一作者的姓加等再加上发表年代，如 Jordan et al. 2015；②如果文中一句话包含多篇参考文献，则在标注参考文献时一般按所引用参考文献发表时间先后顺序或作者姓首字母的音序排列，如 Halevy and Mayak, 1981; Song, 1998; Jordan, 2016。

2. 参考文献列表的具体编排顺序

（1）按作者姓氏字母顺序排列。

（2）按文中参考文献出现的先后顺序编排，即对各参考文献按引用的顺序编排序号。

3. 论文中参考文献选用原则

（1）尽量选用原始文献，杜绝使用二次文献。

（2）尽量选用较新的文献，一般以引用近 3 年发表的相关文章为主，这样更能突出本领域的最新研究进展和自己科研的意义和价值。但是，本领域经典的文献一定要引用。

（3）确保文献正确无误（作者姓名、论文题目、期刊或专著名、期刊的年/卷/期或专著的出版年、出版地、出版社、起止页码等）。

（4）避免过多地特别是不必要地引用作者本人的文献。

（5）尽量引用高水平的文献。

Unit 14 Contributions and publishing skills of English scientific papers

Lesson 58 How to select the periodicals

论文完成后面临的问题就是期刊选择，如何选择期刊是一门学问。选择恰当的期刊是论文能够快速发表的一个重要环节。那么，该如何从众多的期刊中选择比较适合自己文章的期刊呢？对于经验丰富的科研工作者，一般在撰写文章时心中就已经有了安排。而对于大学生和研究生，由于缺乏该方面的经验，通常会觉得无从下手。本课主要介绍选择期刊投稿的一些方法及主要事项，以便帮助投稿者选择出合适的期刊。

一、期刊选择方法

（1）对自己的文章进行自评，即从创新性和研究意义及技术难度等方面进行大致评估，从而为期刊的选择提供指导。

（2）自己所写文章的研究领域要与期刊所包括的领域相一致。投稿前查一下和自己研究相近的论文都在哪些期刊上发表，然后结合自己论文质量进行选择，这样就可以做到有的放矢。从期刊名也可以初步筛选哪些期刊适合自己的文章。然后找到期刊主页，查看其“目标和范围”（aim and scope），以确定稿件是否符合。如果论文发表不着急的话，可以先选择级别较高的期刊投稿，就算被拒，也可以收到很好的关于论文的建议，然后根据编辑部和评审者提出的建议进行详细的修改后再改投。在这里需要补充一点，期刊的影响因子和分区的级别代表了该期刊在该领域中的地位，但每个期刊的影响因子和分区并不

是固定不变的。

（3）期刊的周期和每年刊载量。期刊的周期一般有：周刊、半月刊、月刊、双月刊、季刊、半年刊和年刊。一般来讲周期短的期刊需要的文章数量大，相对来说刊载多的期刊发表的机会可能多一些。

（4）明确所选 SCI 期刊的审稿周期。一篇文章从投稿到正式发表，不同的期刊效率不同。如果你的文章需要在短期内见刊，那就需要选择一个审稿周期短的期刊。那么，如何了解不同期刊的审稿周期呢？可以从近期该期刊发表论文的首页了解该信息。

范例：

若想投 Scientific reports，那么从图 14.1 中圆圈标注的地方可以得知，这篇论文是 2015 年 8 月 14 日被该期刊收到，2015 年 11 月 03 日接收，2015 年 12 月 10 日出版。根据这些信息可以判断一篇文章在该期刊上发表大概需要多长时间。

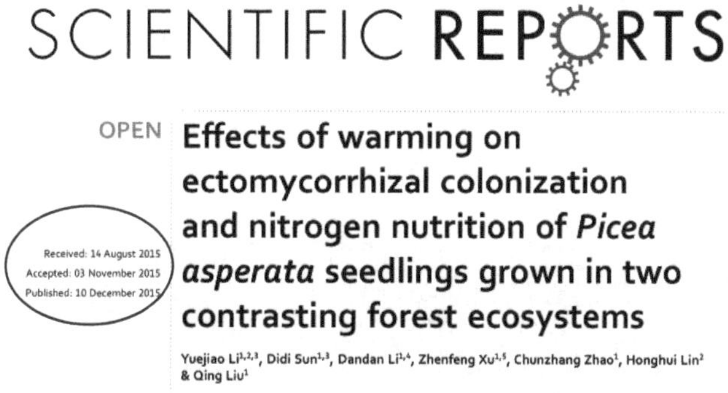

SCIENTIFIC REPORTS

OPEN

Received: 14 August 2015
Accepted: 03 November 2015
Published: 10 December 2015

Effects of warming on ectomycorrhizal colonization and nitrogen nutrition of *Picea asperata* seedlings grown in two contrasting forest ecosystems

Yuejiao Li[1,2,3], Didi Sun[1,3], Dandan Li[1,4], Zhenfeng Xu[1,5], Chunzhang Zhao[1], Honghui Lin[2] & Qing Liu[1]

图 14.1

二、期刊选择注意事项

（1）注意所选英文期刊是否收费。发表英文论文费用一般包括版面费和彩色图片印刷费。期刊收费情况一般在期刊主页中“Author guidelines”中有介绍。SCI 数据库中多数期刊是免费的，但如果作者选择论文开放发表或者论文中的图要求彩色打印则需要缴纳费用。另外，如果作者索要单篇论文打印本也需要缴费。如果没有经费资助的话尽量选择不收费的期刊，因为一般国外期刊收费还是很高的。

（2）不同刊物有不同的固定格式和版式特点，期刊选定后要根据期刊要求对论文进行格式上的修改，尤其是参考文献格式。避免因自己的论文格式与所投刊物要求不相符而被退稿，耽误论文发表。

（3）谨防上当受骗，投 SCI 时注意 ISSN 号和期刊主页上的影响因子。因为有些期刊名称非常相似，甚至有些期刊名称完全相同。例如：图 14.2 和图 14.3 是两个名称完全相同的期刊，但图 14.3 中的期刊是被 SCI 收录的，图 14.2 中的期刊不被 SCI 收录。笔者就被这个期刊欺骗过。在此特意指出，希望后来者不要上当受骗。

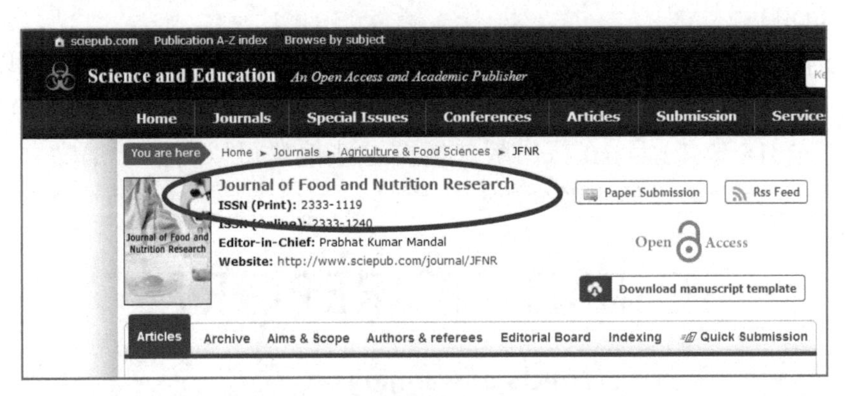

图 14.2

图 14.3

Lesson 59 Submission process and modification process

期刊选择好后，一定要按照所选期刊格式要求对自己的论文从头到尾进行详细的修改，否则会延长发表过程。本课主要从三方面讲解投稿过程中需要做的准备和注意事项。

一、投稿前的准备

1. 申请账号

大多数外文期刊都要求从网上在线投递，只有很少一部分期刊要求通过 e-mail 投稿。下面以 Physiologia Plantarum 期刊为例简单介绍一下账号申请。

（1）打开所选期刊主页（图 14.4）点击“Submit an Article”

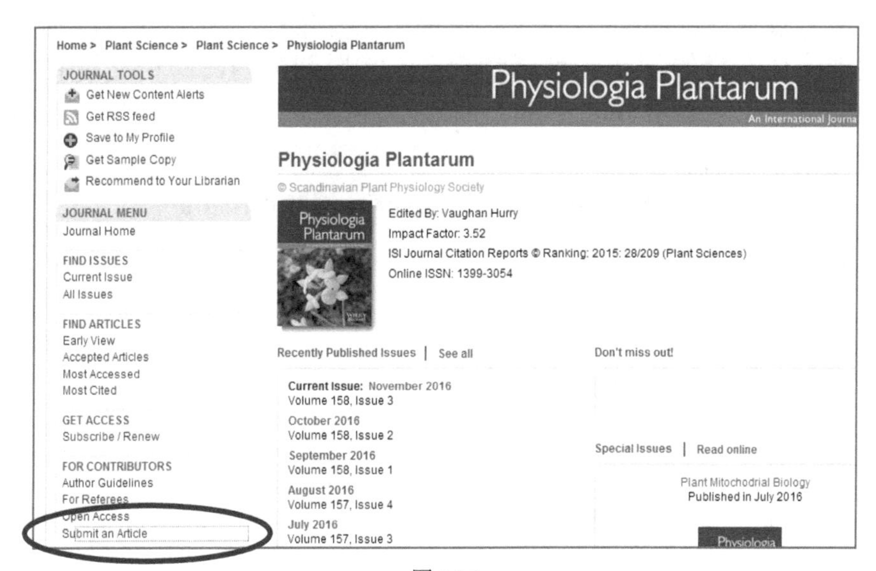

图 14.4

（2）如果已有账号可以直接登录进行投稿，如果没有则点击图 14.5 中的“Register here”进行注册，获得账号和密码，然后登录（Log in）

就可以按要求投稿。

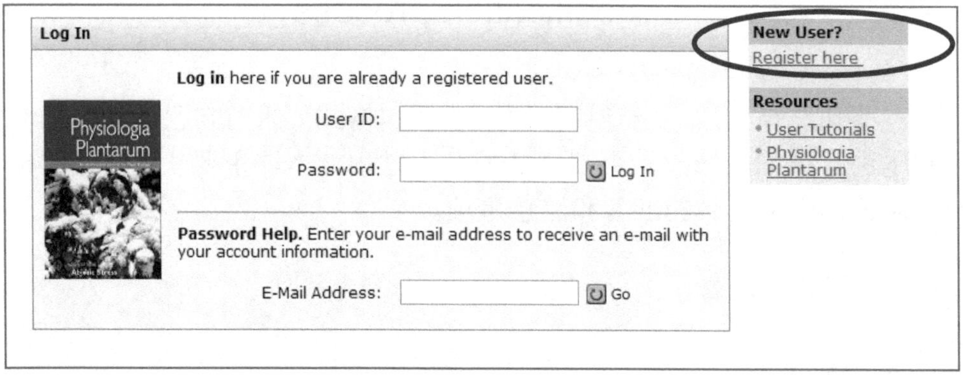

Log In

Log in here if you are already a registered user.

Physiologia Plantarum

User ID:

Password: Log In

Password Help. Enter your e-mail address to receive an e-mail with your account information.

E-Mail Address: Go

New User?

Register here

Resources

- User Tutorials
- Physiologia Plantarum

图 14.5

2. 投稿中需要准备的文件

论文投递过程中，除了需要完整的论文外，一般还需要准备以下文件：

（1）投稿信（Cover letter）。投稿信主要包括三方面内容：①简明扼要地描述文章的创新性和研究意义；②原创声明，即论文没有在其他刊物出版；③论文作者之间没有利益冲突。具体写作请参考下面的范例。

Cover Letter for Manuscript Submission

Type of Contribution:

Title:

*Author Name：

Author University/Organization：

Address：

Date：

* If there are more than one author, enlist them in a serial order with the above mentioned details.

Dear Editor-in-Chief,

I/We am/are writing to submit my/our manuscript entitled "<title>" for consideration for publication in the journal of "<journal name>".

"<Describe the impact and application of your manuscript>" As my/our findings could be applied in XXX way, they are likely to be of great interest to the vision scientists, researchers, clinicians, and students who read your journal on these topics "<specify the topics>".

This manuscript describes novel work and is not under consideration for publication/published by any other journal. All author(s) have approved the manuscript and this submission.

The author(s) certify that there is no conflict of interest with any financial/research/ academic organization, with regards to the content/ research work discussed in the manuscript. "<If any conflict of interest exists please provide the details to the Editorial Office>".

Thank you for receiving my/our manuscript and considering it for review. Kindly contact me/us at the following address for any future correspondence.

Best Regards,
*Corresponding Author's Name
Affiliation
Email Address
Tel. No.:
Fax:

（2）研究创新点（Research Highlights）。研究创新点就是你的研究亮点（创新性或新发现），一般要提供3～5条，每条不超过85个字符（包括空格），目前，好多杂志都要求，主要是给评审者看，最终不出现在论文中。

范例：

Research Highlights

Nutrient and active ingredient contents in flower depended on floral development.

UV-B increased nutrient and active ingredient contents in four stages of flowers.

The best harvest stage of flowers was between the bud and young flower stages.

（3）审稿专家名单（Reviewers List）。大多数期刊需要提供3～5

名论文评审专家，主要包括评审专家的姓名、单位、地址、邮箱和选择的原因。目前期刊评阅人主要来自三方面：①杂志编委会成员；②世界同行、专家学者（编辑部通过关键词检索到类似文章，然后联系文章通信作者）；③作者提供的审稿人。通常编辑部如果能够找到合适的审稿人，就不会采纳作者提供的审稿人。然而当编辑部找不到合适的审稿人时通常会考虑作者所提供的评审人。因此，推荐审稿人时应注意以下几点：

第一，一定要选择和自己研究方向和研究水平相近的专家。这就是投稿者筛选论文审稿人的主要原因。

第二，注意回避一些喜欢到处攻击，或者做各种评审过于刻薄的专家。一些期刊可以提供一些作者需要回避的专家。如果作者不想让自己的论文送到同领域的某位专家手里，这时候一定要提出来。

第三，可以选择自己论文中所引论文中的通信作者作为审稿人。

第四，推荐专家的 e-mail 一定要正确，而且能够联系上，因为编辑部主要通过 e-mail 与所推荐的审稿人取得联系。

二、论文递交过程

投稿前期工作准备好后，即可登录先前在目标期刊上申请的账号，然后按提示一步一步操作即可。在提交后，会收到一封编辑部的信，告诉作者他们已收到论文，并给出论文编号。接下来就是等待编辑部的消息。在等待过程中可以随时登录查询论文的状态。一般来讲主要有以下几种状态。

1. 论文递交到期刊（Submitted to Journal 或 With editor）

当上传结束后，显示的状态是 Submitted to Journal 或 With editor，这个状态是自然形成的，无须处理。如果在投稿的时候没有选择编辑，就先到主编那里，主编会分派给别的编辑。这个过程一般来讲需要 1 周左右，这时投稿者会收到编辑部告知论文编号的来信。

2. 邀请评审专家（Reviewer(s) invited）

说明编辑已接手处理，正在邀请审稿人中。该过程有时会持续很

长时间。可能因为编辑部没有找到合适的审稿人。如果超过 1 个月，建议作者主动给编辑部写信询问一下情况。

3. 评审中（Under review）

说明审稿人已接受审稿，正在审稿中，该过程大约需要 1～2 个月，因为一般期刊通常限定审稿人审稿时间为 1 个月左右。

4. 审稿完成（Required review completed）

表示专家审稿结束，接下来等待编辑处理，该过程一般需要 1～2 周的时间。

5. 最后决定（Decision, Awaiting final decision,Decision in processing）

在该过程，编辑部会充分考虑评审专家的意见来给出结果，其结果主要有以下几种情况：

（1）小修/大修（Minor revision/Major revision）。只要按照审稿人和编辑部提出的意见认真修改，一般都会接收。编辑部会给出具体的修改时间，一般大约为 4～8 周，作者尽量要在截止日前递交修改稿。如果到期了还没有修改好一定要和编辑部进行沟通。

范例 1：

Editorial decision: We have received the reports from our advisors on your manuscript which you submitted to Plant Growth Regulation. Based on the advice received, I feel that your manuscript could be reconsidered for publication should you be prepared to incorporate major revisions. When preparing your revised manuscript, you are asked to carefully consider the reviewer comments which are attached. We look forward to receiving your revised manuscript within eight weeks.

（2）修改后重投（Reject with an invitation to resubmit）。这种结果要求作者修改后再递交，论文将可能作为新的稿件重新处理。

范例 2：

Editorial decision: Reject with an invitation to resubmit. The manuscript appears to include a significant body of potentially worthy results. Both reviewers have raised a number of significant concerns with regard to the content, organization, format and references in the paper and I therefore do not accept your manuscript for publication in its present form. I would be willing, however, to reconsider a properly rewritten and re-submitted manuscript improved according to all the constructive and extensive suggestions made by the reviewers. A re-submitted manuscript will be subject to a second round of review.

（3）拒稿 (Rejected)。这种结果明确表达了编辑的意见，期刊拒绝的原因可能是本论文不适合这个期刊，所以看到这种结果，不要气馁。应该好好选择适合自己论文内容和水平的期刊进行改投。

范例 3：

Editorial decision: I have received the decision from the Editor on your manuscript. This paper presents some interesting results. Unfortunately, it is outside the scope of New Forests; it does not deal with the regeneration phase of forestry.

三、修改稿准备过程

（1）论文要根据编辑部和评审专家提出的意见认真修改。在修改过程中对于中肯意见，一定要接受；对于评审人不正确的意见，一定要进行解释说明；对于无法解决的，也如实说明，表示可以作为下一步的研究工作。另外，要将论文修改处用红色字体显示出来。

（2）准备 Response to reviewer 信件。该信件非常重要，而且也是递交修改稿件时必需的文件。Response to reviewer 信件主要包括以下几方面内容：修改论文的名称、论文的编号、对评审专家和编辑部的感谢及对评审专家和编辑部提出意见的逐条回答等。

范例：

Editor
Journal of Photochemistry and Photobiology B: Biology

Dec 28, 2015

Dear Carpentier,

Thank you very much for your letter on Dec 17, 2015, and comments, with regard to our manuscript "Effects of enhanced UV-B radiation and developmental stages on biochemical traits and quality in postharvest chrysanthemum flowers (JPHOTOBIOL_2015_108)". At present, the title has been changed to "Effects of enhanced UV-B radiation on the nutritional and active ingredient contents during the floral development of medicinal chrysanthemum". We have revised our manuscript according to editor's and reviewers' comments. We believe that the revised manuscript has been improved satisfactorily and hope it will be accepted for publication in Journal of Photochemistry and Photobiology B: Biology. The changes in revised paper were marked with red colour text.

Our responses to reviewers' comments are as follows:

Reviewers' comments

Reviewer #1

1. The authors determined a lot of indices. Therefore, it needs to be more widely discussed about the importance of this indices and UV-B impact on them in the introduction.

Yes, I added some in the introduction of the revised paper, and discussed UV-B impact on them in the discussions.

2. In the Materials and Methods it needs to describe clearly from fresh or dry mass all analysis was made.

Yes, I added it in the revised paper.

Reviewer #2

1. The manuscript (JPHOTOBIOL_2015_108), "Effects of enhanced UV-B radiation and developmental stages on biochemical traits and quality in postharvest chrysanthemum flowers", is a revision. This reviewer believes that the manuscript must be revised before being considered for publication in Journal of Photochemistry and Photobiology, B: Biology.

Yes, I carefully revised it in the revised paper.

2. Title is confusing and should be "Effects of enhanced UV-B radiation on the nutrient and active ingredient contents during the floral development of medicinal chrysanthemum"

Yes, I changed it in the revised paper.

3. Abstract: line 33, remove H_2O_2

Because, H_2O_2 has appeared two times in the abstract, the abbreviation was used in second time.

Thanks again for your kind help and comments and suggestions.

With best regards,

Sincerely yours,

Xiao-Qin Yao

当论文修改稿和对审稿专家所提意见的答复（Response to reviewer）信件准备好后，就可以登录账号递交，切记这时候不要把修改稿作为一个新稿件递交。最后，作者将会收到接收（Accepting）信，有的需要再经过几次的修改才被接受。所以，一篇论文从开始写作，到最后的发表都要经过一个很长的过程。在校大学生若想论文在毕业前能够发表，一定要合理安排好时间。

References

Part I

Engelking LR. 2015. Textbook of Veterinary Physiological Chemistry (Third Edition). USA, San Diego: Academic Press.

https://en.wikipedia.org/wiki/.

https://www.boundless.com/biology/.

Part II

Aarabi F, Kusajima M, Tohge T, et al. 2016. Sulfur deficiency-induced repressor proteins optimize glucosinolate biosynthesis in plants. Science Advances, 2: 1-17.

Albuquerque N, Guo K, Wilkinson A, et al. 2016. Dogs recognize dog and human emotions. Biology Letters, 12: 1-5.

Blaser MJ. 2016. Antibiotic use and its consequences for the normal microbiome. Science, 352: 544-545.

Chen A. 2014. Ninety-nine percent of the ocean's plastic is missing. Science News.

Chu C, Bartlett M, Wang Y et al. 2016. Does climate directly influence NPP globally? Global Change Biology, 22: 12-24.

Dell' Amore C. 2015. Why Do Zebras Have Stripes. New Science.

Dong YS, Fu CH, Su P, et al. 2016. Mechanisms and effective control of physiological browning phenomena in plant cell cutures. Physiologia Plantarum, 156: 13-28.

Henriksson A, Wardle DA, Trygg J, et al. 2016. Strong invaders are strong defenders–implications for the resistance of invaded communities. Ecology Letters, 19: 487-494.

Inouye DW. 2015. The next century of ecology. Science, 349: 565.

Jansson M, Karlsson J, Jonsson A, et al. 2012. Carbon dioxide supersaturation promotes primary production in lakes. Ecology Letters, 15: 527-532.

Joel. 2016. How Earth's oldest animals were fossilized. Science News.

Katherine Harmon Courage. 2016. Your poor diet might hurt your grandchildren's guts. Science News.

Kintisch E. 2015. Earth's lakes are warming faster than its air. Science News.

Leslie M. 2015. Gut microbes give anticancer treatments a boost. Science News.

Leslie M. 2016. Suicide of aging cells prolongs life span in mice. Science News.

Lipka E, Herrmann A, Mueller S. 2015. Mechanisms of plant cell division. Wiley Interdisciplinary Reviews: Developmental Biology, 4: 391-405.

McCauley DJ, Pinsky ML, Palumbi SR, et al. 2015. Marine defaunation: Animal loss in the global ocean. Science, 347: 247.

Moeendarbary E, Harris AR. 2014. Cell mechanics: principles, practices, and prospects. Wires Systems Biology and Medicine, 6: 371-388.

Moran MA. 2015. The global ocean microbiome. Science, 360: 1330-1335.

Morell V. 2016. Plants can gamble, according to study. Science News.

Pennisi. 2016. Tiny DNA tweaks made snakes legless. Science News.

Schlebusch CM, Gattepaille LM, Engström K. 2015. Human adaptation to arsenic-rich environments. Molecular Biology and Evolution, 32: 1544-1555.

Science News Staff. 2016. Nobel honors discoveries on how cells eat themselves. Science News.

Servellen SV, Oba I. 2014. Stem cell research: Trends in and perspectives on the evolving international landscape. Research Trends, 36: 6-10.

Suzuki N, Rivero RM, Shulaev V, et al. 2014. Abiotic and biotic stress combinations. New Phytologist, 203: 32-43.

University of Exeter. 2016. Tiny microbe turns tropical butterfly into male killer. Bioscience.

Vogel G. 2016. Zika virus kills developing brain cells. Science News.

Zeuss D, Brandl R, Brändle M, et al. 2014. Global warming favours light-coloured insects in Europe. Nature Communications, 5: 1-10.

Zobel RW. 2016. Fine roots-functional definition expanded to crop species. New Phytologist, 212: 310-312.

Part III

杨冬艳. 2009. 浅谈大学生电子文献检索中的常见问题. 图书馆论坛, 29: 147-148.

Águila Ruiz-Sola M, Coman D, Beck G, et al. 2014. Arabidopsis GERANYLGERANYL DIPHOSPHATE SYNTHASE 11 is a hub isozyme required for the production of most photosynthesis-related isoprenoids. New Phytologist, 209, (1): 252-264.

Li YJ, Sun DD, Li DD, et al. 2015. Effects of warming on ectomycorrhizal colonization and nitrogen nutrition of Picea asperata seedlings grown in two contrasting forest ecosystems. Scientific Reports. 5: 1-10.

Yao XQ, Chu JZ, He XL, et al. 2013. Effects of selenium on agronomical characters of winter wheat exposed to enhanced ultraviolet-B. Ecotoxicology and Environmental Safety, 92: 320-326.

Yao XQ, Chu JZ, He XL, et al. 2014. Grain yield, starch, protein, and nutritional element concentrations of winter wheat exposed to enhanced UV-B during

different growth stages. Journal of Cereal Science, 60: 31-36.

Yao XQ, Chu JZ, Ma CH, et al. 2015. Biochemical traits and proteomic changes in postharvest flowers of medicinal chrysanthemum exposed to enhanced UV-B radiation. Journal of Photochemistry and Photobiology B: Biology, 149: 272-279.

Yao XQ, Liu Q. 2007. Changes in photosynthesis and antioxidant defenses of Picea asperata seedlings to enhanced ultraviolet-B and to nitrogen supply, Physiologia Plantarum, 129: 364-374.

Appendix

Appendix Ⅰ Common biology core vocabulary

Glossary related to cell

actin 肌动蛋白
apoptosis 细胞凋亡
apoptotic body 凋亡小体
autophagy 细胞自噬
genome 基因组
biomembrane 生物膜
cell biology 细胞生物学
cell differentiation 细胞分化
cell junction 细胞连接
cell proliferation 细胞增殖
cell wall 细胞壁
cellulose 纤维素
centriole 中心粒
centromere 着丝粒
centrosome 中心体
chloroplast 叶绿体
chromosome 染色体
chromosome complement 染色体组
cytoplasm 细胞质
cytoskeleton 细胞骨架
cytosol 胞质溶液
endoplasmic reticulum 内质网
smooth endoplasmic reticulum 光面内质网
rough endoplasmic reticulum 糙面内质网
golgi body 高尔基体
eukaryotic cell, eukaryocyte 真核细胞
glucose transporter 葡萄糖转运蛋白
glycerophosphatide 甘油磷脂
glycosaminoglycan 糖胺聚糖
glycosyl-transferase 糖基转移酶
golgi complex 高尔基复合体
histone 组蛋白
integral protein 整合蛋白
karyotype 核型
lignin 木质素
lysosome 溶酶体
meiosis 减数分裂
microbody 微体
microfilament 微丝
microtubule 微管
mitochondria 线粒体
mitosis 有丝分裂
nuclear membrane 核膜
nucleolus, nucleoli （复）核仁
nucleolus 核仁
nucleosome 核小体

nucleus 细胞核
organelle 细胞器
oxidative phosphorylation 氧化磷酸化
pinocytosis 胞饮作用
plasmodesma 胞间连丝
plasmolemma 质膜
plasmolysis cytoplasm 质壁分离
plastid 质体
polarization 极化
Polyribosome 多核糖体
programmed cell death 程序性细胞死亡
prokaryotic cell, prokaryocyte 原核细胞
protoplasm 原生质
receptor 受体
respiratory chain 呼吸链
ribosome 核糖体
signal molecule 信号分子
signal transduction 信号转导
signaling pathway 信号通路
spindle 纺锤体
stem cell 干细胞
sterol 固醇
telomerase 端粒酶
thylakoid 类囊体
transmembrane protein 跨膜蛋白
vacuole 液泡

Glossary related to microorganisms

aerobic respiration 有氧呼吸
agar 琼脂
algae 藻类
anaerobic respiration 无氧呼吸
antibiotic 抗生素
antibody 抗体
antigen 抗原
antiport 逆向运输
archaea 古生菌
autotrophys 自养型
bacterium 细菌
bacteriolysis 溶菌剂
binary fission 二分裂
biomass 生物量
budding 芽殖
capsid 衣壳
capsule 荚膜
carrier 载体
cell mediated immunity 细胞免疫
chitin 几丁质
cilia 纤毛
colony 菌落
conjugation 接合作用
continuous culture 连续培养
culture dish 培养皿
culture medium 培养基
culture plate 培养平板
culture 培养物
cyst 孢囊
diffusion 扩散
disinfectant 消毒剂
dormant 休眠
exospore 外生孢子
fermentation 发酵
fission 裂殖
fungus 真菌
gene mutation 基因突变
genotype 基因型
glycoprotein 糖蛋白

heterotrophs 异养型
hyphae 菌丝
immunity 免疫
immunoglobulin 免疫球蛋白
inflammatory 炎症
lysozyme 溶菌酶
metabolism 代谢
microbiology 微生物学
mold 霉菌
mycelium 菌丝体
mycorrhiza 菌根
pathogenic microorganism 病原微生物
peptidoglycan 肽聚糖
phage 噬菌体
phenotype 表型
pinocytosis 胞饮作用
plasmid 质粒
protozoan 原生动物
redox potential 氧化还原电位
spore 孢子
symport 同向运输
trace element 微量元素
transduction 转导
transformation 转化
transition 转换
viroid 类病毒
virus 病毒
yeast 酵母菌

Glossary related to animals

air sac 气囊
allantois 尿囊
antennae 触角
appendage 附肢
axial skeleton 中轴骨骼
bilateral symmetry 两侧对称
bill 喙
blastocoel 囊胚腔
blastoderm 囊胚层
blastomere 分裂球
blastula 囊胚
cap cell 帽细胞
cardiac muscle 心肌
cartilage tissue 软骨组织
cleavage 卵裂
cloacal pore 泄殖孔
cnidoblast 刺细胞
coelenteron 腔肠
colony 群体
connective tissue 结缔组织
contractile vacuole 伸缩炮
delamination 分层
dendron 树突
dense connective tissue 致密结缔组织
dermatome 生皮节
deuterostome 后口动物
dioecism 雌雄异体
dual respiration 双重呼吸
ecdysis 蜕皮
epithelial tissue 上皮组织
excurrent siphon 出水孔
feather epithelium 羽状上皮
feather 羽毛
fertilization 受精
ganoid scale 硬鳞
gestation 妊娠
gill slits 腮裂
glandular epithelium 腺上皮
inversion 逆转
involution 内转

loose connective tissue　疏松结缔组织
mesoderm　中胚层
metamerism　分节现象
metazoa　后生动物
migration　洄游
monoecism　雌雄同体
moulting hormone　蜕皮激素
muscular tissue　肌肉组织
nervous tissue　神经组织
nervus ganglion　神经节
neuron　神经元
obliquely striated muscle　斜纹肌
organ　器官
osmotrophy　渗透营养
osseous tissue　骨组织
peritoneum　体腔膜
phagocytosis　吞噬作用
phylogenetic tree　种系发生树
phylogeny　种系发生
placenta　胎盘
protostome　原口动物
radial symmetry　辐射对称
regeneration　再生
septum　隔膜
smooth muscle　平滑肌
tissue　组织
total cleavage　完全卵裂
vivipary　胎生
wing　翼
zoology　动物学

Glossary related to plant

alternation of generations　世代交替
annual ring　年轮
anther　花药
apical dominance　顶端优势
assimilation　同化作用
autotrophic plant　自养植物
auxin　生长素
bisexual flower　两性花
carpel　心皮
casparian strip　凯氏带
climbing stem　攀缘茎
collenchyma　厚角组织
companion cell　伴胞
corpus callosum　胼胝体
cortex　皮层
creeping stem　匍匐茎
cuticle　角质层
cutting　扦插
cytokinin　细胞分裂素
dicot　双子叶植物
endosperm　胚乳
epidermis　表皮
fascicular cambium　束中形成层
gametophyte　配子体
grafting　嫁接
guard cell　保卫细胞
heterotrophic plant　异养植物
hydrophyte　水生植物
intercellular layer　胞间层
interfascicular cambium　束间形成层
layering　压条
mature tissue　成熟组织
meristematic zone　分生区
meristem　分生组织
mesophyll　叶肉
monocot　单子叶植物
node　节
ovary　子房

ovule　胚珠
palisade tissue　栅栏组织
parasite　寄生植物
periderm　周皮
petal　花瓣
petiole　叶柄
phloem　韧皮部
photosystem　光系统
phototropism　向光性
pistil　雌蕊
pit　纹孔
plasmodesma　胞间连丝
plumule　胚芽
pollination　传粉
primary growth　初生生长
primary wall　初生壁
sclereid　石细胞
secondary wall　次生壁
sepal　萼片
sieve tube　筛管
spongy tissue　海绵组织
sporophyte　孢子体
stamen　雄蕊
stem　茎
stigma　柱头
style　花柱
tiller　分蘖
tracheid　管胞
vascular cylinder　维管柱
vein 叶脉
woody plant　木本植物
xylem　木质部
zygote　合子

Glossary related to Ecology

abiotic environment　非生物环境
abundance　多度，丰富度
acclimation　驯化
age structure　年龄结构
allelopathy　他感作用
assimilation efficiency　同化效率
autotroph　自养生物
biological clock　生物钟
biodiversity　生物多样性
biomass　生物量
biome　生物群系
biotic environment　生物环境
carnivore　食肉动物
carrying capacity　容纳量
community　群落
companion species　伴生种
competition　竞争
constructive species　建群种
coverage　盖度
cryptic coloration　保护色
daily rhythm　昼夜节律
decomposer　分解者
density dependent factor　密度制约因子
density independent factor　非密度制约因子
desert　荒漠
detrital food chain　碎屑食物链
dominant species　优势种
dormancy　休眠
ecolog　生态学
ecological amplitude　生态幅
ecological factor　生态因子
ecological invasion　生态入侵

ecosystem　生态系统
ecotone　群落交错区
emigration　迁出
eoevolution　协同进化
feedback mechanism　反馈机制
fitness　适合度
food chain　食物链
food web　食物网
frequency　频度
gene flow　基因流
generation　世代
genetic drift　遗传漂变
habitat　生境
herbivores　食草动物
hibernation　冬眠
homeotherm　常温动物
humus　腐殖质
hygrophyte　湿生植物
immigration　迁入
individuals　个体
interspecific competition　种间竞争
intraspecific relationship　种内关系
key factor　关键因子
killing factor　致死因子
landscape　景观
law of tolerance　耐受性定律
life expectancy　生命期望
longevity　寿命
marcroenvironment　大环境
mating system　婚配制度
microenvironment　小环境
mortality　死亡率
natality　出生率
natural selection　自然选择
parasitism　寄生
patch　斑块
poikilotherm　变温动物
population　种群
precipitation　降雨量
predation　捕食
relative density　相对密度
reproductive isolating mechanism　繁殖隔离机制
simpson's diversity index　辛普森多样性指数
social hierarchy　社会等级
species richness　物种丰富度
succession　演替
survivorship curve　存活曲线
tundra　冻原
vernalization　春化

Appendix II Research paper example

Physiologia Plantarum 129: 364–374. 2007

Changes in photosynthesis and antioxidant defenses of *Picea asperata* seedlings to enhanced ultraviolet-B and to nitrogen supply

Xiaoqin Yao[a,b] and Qing Liu[a,*]

[a]Chengdu Institute of Biology, Chinese Academy of Sciences, Post Box 416, Chengdu 610041, China
[b]Graduate School of the Chinese Academy of Sciences, Beijing 100039, China

Correspondence
*Corresponding author,
e-mail: liuqing@cib.ac.cn

Received 30 May 2006; revised 25 July 2006

doi: 10.1111/j.1399-3054.2006.00815.x

The paper mainly studied the effects of ultraviolet-B (UV-B) radiation, nitrogen, and their combination on photosynthesis and antioxidant defenses of *Picea asperata* seedlings. The experimental design included two levels of UV-B treatments (ambient UV-B, 11.02 KJ m^{-2} day^{-1}; enhanced UV-B, 14.33 KJ m^{-2} day^{-1}) and two nitrogen levels (0; 20 g m^{-2} a^{-1} N) to determine whether the adverse effects of UV-B are eased by supplemental nitrogen. Enhanced UV-B significantly inhibited plant growth, net photosynthetic rate (A), stomatal conductance to water vapor (Gs), transpiration rate and photosynthetic pigment, and increased intercellular CO_2 concentration, UV-B absorbing compounds, proline content, malondialdehyde (MDA) content, and activity of antioxidant enzymes (peroxidase (POD), superoxide dimutase, and glutathione reductase). Enhanced UV-B also reduced needle DW and increased hydrogen peroxide (H_2O_2) content and the rate of superoxide radical (O_2^-) production only under supplemental nitrogen. On the other hand, supplemental nitrogen increased plant growth, A, Gs, chlorophyll content and activity of antioxidant enzymes (POD, ascorbate peroxidase, and catalase), and reduced MDA content, H_2O_2 content, and the rate of O_2^- production only under ambient UV-B, whereas supplemental nitrogen reduced activity of antioxidant enzymes under enhanced UV-B. Carotenoids content, proline content, and UV-B absorbing compounds increased under supplemental nitrogen. Moreover, significant UV-B × nitrogen interaction was found on plant height, basal diameter, A, chlorophyll *a*, activity of antioxidant enzymes, H_2O_2, MDA, and proline content. These results implied that supplemental nitrogen was favorable for photosynthesis and antioxidant defenses of *P. asperata* seedlings under ambient UV-B. However, supplemental nitrogen made the plants more sensitive to enhanced UV-B, although some antioxidant indexes increased.

Abbreviations – A, net photosynthetic effect; APX, ascorbate peroxidase; C, ambient UV-B without supplemental nitrogen; Car, carotenoids, CAT, catalase; Chl *a*, chlorophyll *a*; Chl *b*, chlorophyll *b*; Chl (*a* + *b*), total chlorophyll; Ci, intercellular CO_2 concentration; E, transpiration rate; GR, glutathione reductase; H_2O_2, hydrogen peroxide; MDA, malondialdehyde; N, ambient UV-B with supplemental nitrogen; NADPH, nicotinamide adenine dinucleotide phosphate; NBT, nitroblue tetrazolium; N × UV-B, interaction effect of UV-B and nitrogen; O_2^-, superoxide radical; POD, peroxidase; PPFD, photosynthetic photon flux density; ROS, reactive oxygen species; SOD, superoxide dimutase; TBA, thiobarbituric acid; TCA, trichloroacetic acid; UV, ultraviolet; UV-B, enhanced UV-B without supplemental nitrogen; UV-B + N, enhanced UV-B with supplemental nitrogen.

Introduction

Atmospheric ozone remains depleted and the annual average ozone loss is approximately 3% globally (Executive summary 2003). Researches have shown that enhanced ultraviolet-B (UV-B) reaching the surface of the earth has very many adverse impacts on plant growth (Jordan 1996, 2002, Jansen 2002). When plants are exposed to UV-B stress, they could induce some protective mechanisms. For example, the increases in UV-B absorbing compounds, proline content, and activity of antioxidant enzymes have been reported (Baumbusch et al. 1998, Prochazkova et al. 2001, Saradhi et al. 1995).

In addition to UV-B radiation, human activities have significantly altered the global nitrogen cycle, with the development of industry and agriculture. More and more nitrogen will be imported into the terrestrial ecosystems through nitrogen deposition. In the European livestock and industrialized areas, nitrogen deposition was more than 25 kg hm^{-2} a^{-1} N (Binkley et al. 2000). In the Northeastern United States, the current nitrogen deposition was more 10–20 times than nitrogen in background (Magill et al. 1997). At present, China has been one of three high-nitrogen deposition regions (Li et al. 2003). Nitrogen is the mineral nutrient needed in largest amounts by plants and it is usually also the limiting factor for plant growth in terrestrial ecosystems (Vitousek and Howarth 1991), particularly in tundra, boreal as well as alpine ecosystems (Xu et al. 2003). At the same time, nitrogen is also an important constituent of photosynthetic apparatus (Correia et al. 2005). Maximum photosynthetic capacity is strongly regulated by leaf nitrogen concentration (Field and Mooney 1986). In contrast to UV-B radiation, supplemental nitrogen improved growth and net photosynthesis of plant (Nakaji et al. 2001, Keski-Saari and Julkunen-Tiitto 2003) and reduce production of free radicals in plants (Ramalho et al. 1998).

UV-B radiation and nitrogen are expected to increase simultaneously with future changes in global climate. Nitrogen can affect UV-B response in plants (Correia et al. 2005, Pinto et al. 1999). Previous studies have mainly focused on crop and herb plants, although forests account for over two-thirds of global net primary productivity (NPP), compared with about 11% for agricultural land (Barnes et al. 1998). However, only limited papers have been reported on the combined effects of nitrogen and UV-B radiation on woody plants (De La Rose et al. 2001, 2003, Lavola et al. 2003).

Picea asperata is a key species in the southeast of the Qinghai-Tibetan Plateau of China and widely used in reforestation programs at present (Liu 2002). The paper mainly studies the short-term influence of enhanced UV-B radiation and supplemental nitrogen on photosynthesis and antioxidant defenses of *P. asperata* seedlings under semi-field condition. This will be helpful for understanding of the combined effects on conifer tree species and development of improved plant tolerance toward stressful environmental factors. On the basis of previous study in other species, we hypothesized that (1) both UV-B and nitrogen would affect photosynthesis and antioxidant defenses of *P. asperata* seedlings; and (2) supplemental nitrogen modifies the adverse effects of UV-B on the conifer plants, in order to better understand the responses of woody plant to both enhanced UV-B and to supplemental nitrogen in future.

Materials and methods

Plant material and experiment design

The experiment was conducted in open semi-field condition from April 15 to October 15, 2005 in Maoxian Ecological Station of Chinese Academy of Sciences, Sichuan province, China (31°41′ N, 103°53′ E, 1820 m a.s.l.). Four-year-old *P. asperata* seedlings were from a local nursery. The plant height, basal diameter and whole-plant FW at the beginning of the experiment were 15.38 ± 0.48 cm, 6.52 ± 0.35 mm, and 7.52 ± 0.43 g, respectively. Seedlings were transplanted into plastic pots (25-cm diameter and 35-cm depth) with a 12-h photoperiod and a daily average 1200 μmol m^{-2} s^{-1} photosynthetic photon flux density (PPFD), one seedling per pot. The substrate used for growing the seedlings was sieved topsoil from a spruce-forest. In a preliminary experiment, the plastic pots did not affect growth of seedling root during a 2-year growth period.

The experiment consisted of four treatments in the paper: (1) ambient UV-B without supplemental nitrogen (control, C); (2) ambient UV-B with supplemental nitrogen (N); (3) enhanced UV-B without supplemental nitrogen (UV-B); and (4) enhanced UV-B with supplemental nitrogen (UV-B + N). Each treatment has three blocks and each block have 10 pots. The pots within blocks were rotated approximately every 20 days.

UV-B treatments and nitrogen treatments

Supplementary UV-B was supplied by UV-B fluorescent lamps (Beijing Electronic Resource Institute, Beijing, China) mounted in metal frames with minimum shading. The distance from the lamps to the top of plant apex was 100 cm and kept constant throughout the experiment. In ambient UV-B frames, UV-B from the lamps was excluded by wrapping the tubes with 0.125-mm polyester film (Chenguang Research Institute of Chemical Industry, Chengdu, China), which transmits UV-A. In enhanced

UV-B frames, lamps were wrapped with 0.10-mm cellulose diacetate film, which transmits both UV-B and UV-A. Vertical polyester curtains were placed between the frames in order to prevent the UV-B radiation from reaching the C seedlings (De La Rose et al. 2003). Films were replaced every 1 week. The lamp duration was modified monthly and replaced in times. The spectral irradiance from the lamps was determined with an Optronics Model 742 (Optronics Laboratory Inc., Orlando, FL) spectroradiometer. The spectral irradiance was weighted according to the generalized plant action spectrum (Caldwell 1971) and normalized at 300 nm to obtain effective radiation (UV-B_{BE}). The supplemental UV-B_{BE} dose was 3.31 KJ m^{-2} day^{-1} (a 30% difference in ambient UV-B_{BE}) in addition to the effective 11.02 KJ m^{-2} day^{-1} UV-B_{BE} (ambient UV-B_{BE}) from sky. All pots also received natural solar radiation. Seedlings were irradiated for 8 h daily centered on the solar noon.

Nitrogen was added as 9.5 m*M* NH_4NO_3 solution (300 ml) to the potted soil surface every 3 days. The treatment without supplemental nitrogen was watered with 300 ml of water. The nitrogen amount added to the soil was equivalent to 20 gm^{-2} a^{-1} N on the basis of soil surface area. Nitrogen amount was based on the similar studies (Bowden et al. 2004, Nakaji et al. 2001).

Growth parameters

Plant height and basal diameter of six randomly selected seedlings from each treatment were measured and were harvested at the end of the experiment. Seedlings were divided into needle, root, and stem. Root was rinsed free of soil. All the organs were dried at 80°C for 1 week and weighed.

Pigment analysis

Samples of the youngest, fully expanded needles were taken for the determination of chlorophyll content. Needle was grinded in 80% acetone for the determination of chlorophyll and carotenoids (Car). Total chlorophyll [Chl (*a* + *b*)], chlorophyll *a* (Chl *a*), chlorophyll *b* (Chl *b*) and total Car contents were determined according to Lichtenthaler (1987).

Gas exchange

After 60 days of treatment, on a cloudless day. Readings of net photosynthetic rate (A, μmol m^{-2} s^{-1} CO_2), transpiration rate (E, mmol m^{-2} s^{-1}), stomatal conductance to water vapor (Gs, mmol m^{-2} s^{-1}), and intercellular CO_2 concentration (Ci, μmol mol^{-1}) were taken at a saturating PPFD of 1000 ± 50 μmol m^{-2} s^{-1} with a portable photosynthesis system (LI-6400, Lincoln, NE) in the open-circuit mode. Measurements were made around midday.

UV-B absorbing compounds, proline, and malondialdehyde content

UV-B absorbing pigments of the youngest, fully developed needle were extracted from fresh needle material with a MeOH : H_2O : HCl (79 : 20 : 1, v/v/v) solution. Samples were heated in a water bath (90°C) for 1 h. The absorbance at 285 nm was recorded using a scanning spectrophotometer (Unicam UV-330, Thermo Spectronic, Cambridge, UK).

The free proline content was determined according to the method described by Bates et al. (1973). Needles (1 g) were homogenized using a pestle and mortar with 5 ml of sulfosalicylic acid (3% w/v). After centrifugation (5 min at 20 000 *g*), 0.5 ml of the supernatant was incubated at 100°C for 60 min with 0.5 ml of glacial acetic acid and 0.5 ml of ninhydrin reagent. After cooling, 1 ml of toluene was added to the mixture and the absorbance of the chromophore-containing toluene was recorded at 520 nm.

The degree of lipid peroxidation was assessed by malondialdehyde (MDA) content. MDA content was determined by the thiobarbituric acid (TBA) reaction. Needles (0.5 g) were homogenized with 5 ml of 20% (w/v) trichloroacetic acid (TCA). The homogenate was centrifuged at 3500 *g* for 20 min. To 2 ml of the aliquot of the supernatant, 2 ml of 20% TCA containing 0.5% (w/v) TBA and 100 μl of 4% (w/v) butylated hydroxytoluene in ethanol were added. The mixture was heated at 95°C for 30 min and then quickly cooled on ice. The contents were centrifuged at 10 000 *g* for 15 min and the absorbance was measured at 532 nm. The value for non-specific absorption at 600 nm was subtracted. The concentration of MDA was calculated using an extinction coefficient of 155 mM^{-1} cm^{-1}. Results were expressed as μmol g^{-1} FW.

The rate of superoxide radical production and hydrogen peroxide content

The rate of superoxide radical production (O_2^-) was measured as described by Ke et al. (2002), by monitoring the nitrite formation from hydroxylamine in the presence of O_2^-. Needles (0.5 g) were homogenized with 1.5 ml of 65-m*M* potassium phosphate (pH 7.8) and centrifuged at 5000 *g* for 10 min. The incubation mixture contained 0.45 ml of 65-m*M* phosphate buffer (pH 7.8), 0.5 ml of 10 m*M* hydroxylamine hydrochloride, and 0.5 ml of the supernatant. After incubation at 25°C for 20 min, 8.5 m*M* sulfanilamide and 3.5 m*M* α-naphthylamine were added to the incubation mixture. After reaction at 25°C for

20 min, the absorbance in the aqueous solution was read at 530 nm. A standard curve with NO_2^- was used to calculate the production rate of O_2^- from the chemical reaction of O_2^- and hydroxylamine.

Hydrogen peroxide (H_2O_2) content was determined according to Prochazkova et al. (2001). Needle (0.5 g) material was grinded with 5 ml cooled acetone in a cold room (10°C). Mixture was filtered with filter paper followed by the addition of 2 ml titanium reagent and 5 ml ammonium solution to precipitate the titanium–hydrogen peroxide complex. Reaction mixture was centrifuged at 10 000 *g* for 10 min. Precipitate was dissolved in 5 ml of 2-*M* H_2SO_4 and then recentrifuged. Supernatant was read at 415 nm.

Antioxidant enzymes activity

Extracts for the determination of antioxidant enzymes activities were prepared from 1.0 g of fully developed new needles homogenized under ice-cold conditions in 3 ml of extraction buffer, containing 50 m*M* phosphate buffer (pH 7.4), 1 m*M* ethylenediamine-tetraacetic acid (EDTA), 1 g polyvinylpyrrolidone (PVP) and 0.5% (v/v) Triton X-100). The homogenates were centrifuged at 10 000 *g* for 30 min and the supernatant fraction was used for the assays.

Superoxide dimutase activity (SOD, EC 1.15.1.1) was assayed by the inhibition of the photochemical reduction of nitroblue tetrazolium (NBT), as described by Becana et al. (1986). The reaction mixture consisted of 50 μl of enzyme extract and 3.0 ml O_2^- generating mixture solution containing 50 m*M* potassium-phosphate (pH 7.8), 0.1 m*M* Na_2EDT, 13 m*M* methionine, 75 μ*M* NBT, and 16.7 μ*M* riboflavin. Test tubes were shaken and placed 30 cm from light bank consisting of six 15-W fluorescent lamps. The reaction was allowed to run for 10 min and stopped by switching the light off. The reduction in NBT was followed by reading absorbance at 560 nm. Blanks and C were run the same way but without illumination and enzyme, respectively. One unit of SOD was defined as the amount of enzyme, which produced a 50% inhibition of NBT reduction under the assay conditions (Costa et al. 2002).

Catalase activity (CAT, EC 1.11.1.6) was determined in the homogenates by measuring the decrease in absorption at 240 nm in a reaction medium containing 50 m*M* potassium phosphate buffer (pH 7.2), 10 m*M* H_2O_2 and 50 μl enzyme extract. The activity was calculated using the extinction coefficient (40 mM^{-1} cm^{-1}) for H_2O_2.

Ascorbate peroxidase activity (APX, EC 1.11.1.11) was measured in fresh extracts using a reaction mixture containing 50 m*M* potassium phosphate buffer (pH 7.0), 0.1 m*M* H_2O_2, 0.5 m*M* ascorbate, and 0.1 m*M* EDTA. The H_2O_2-dependent oxidation of ascorbate was followed by monitoring the absorbance decrease at 290 nm (extinction coefficient 2.8 mM^{-1} cm^{-1}).

Peroxidase activity (POD, EC 1.11.1.7) was based on the determination of guaiacol oxidation (extinction coefficient 26.6 mM^{-1} cm^{-1}) at 470 nm by H_2O_2. The reaction mixture contained 50 m*M* potassium phosphate buffer (pH 7.0), 20.1 m*M* guaiacol, 12.3 m*M* H_2O_2, and enzyme extract in a 3 ml volume.

Glutathione reductase activity (GR, EC 1.6.4.2) was measured by following the decrease in absorbance at 340 nm (extinction coefficient 6.2 mM^{-1} cm^{-1}) due to nicotinamide adenine dinucleotide phosphate (NADPH) oxidation. The reaction mixture contained 50 μl enzyme extract, 1 m*M* EDTA, 0.5 m*M* glutathion reductase (GSSG), 0.15 m*M* NADPH and 50 m*M* Tris–HCl buffer (pH 7.5) (Costa et al. 2002).

Soluble protein contents were determined as described by Bradford (1976), using bovine serum albumin as a calibration standard.

Statistical analysis

All data were subjected to an analysis of variance that tested the UV-B radiation, nitrogen and UV-B × nitrogen interaction effects, and the significance of the single factors calculated as well as the interaction between the factors calculated. All statistical analyses were performed using the Software Statistical Package for the Social Science (SPSS) version 11.0 (SPSS Inc., Chicago, IL).

Results

The effects of enhanced UV-B and nitrogen supply on growth parameters

Enhanced UV-B significantly decreased plant height, root weight, and total biomass of *P. asperata* seedlings (Table 1). Under supplemental nitrogen, needle DW decreased significantly under enhanced UV-B. On the other hand, supplemental nitrogen significantly increased plant height of *P. asperata* seedlings, whereas basal diameter and total biomass were affected by supplemental nitrogen only under ambient UV-B. Significant interactive effects of UV-B × nitrogen were also detected on plant height and basal diameter.

The effects of enhanced UV-B and nitrogen supply on photosynthetic parameters

Enhanced UV-B markedly reduced A, Gs and E of *P. asperata* seedlings, and increased Ci significantly

Table 1. The effects of enhanced UV-B and supplemental nitrogen on growth parameters of *P. asperata*. Values are the mean ± SE of six replicates in column rows 1–4 and the values in the same column with different letters are significantly different from each other ($P < 0.05$). Significant effects of the two factors as well as of the interaction are indicated in column rows 5–7.

Treatment	Height (cm)	Basal diameter (mm)	Dry mass (g)			
			Root	Stem	Needle	Total
C	18.36 ± 0.06^{b}	7.08 ± 0.26^{b}	$9.59 \pm 0.69^{a,b}$	$5.83 \pm 0.41^{a,b}$	$5.37 \pm 0.28^{a,b}$	20.80 ± 0.27^{b}
N	24.10 ± 0.28^{a}	8.48 ± 0.15^{a}	11.52 ± 0.51^{a}	6.59 ± 0.15^{a}	6.21 ± 0.6^{a}	24.32 ± 0.71^{a}
UV-B	16.46 ± 0.46^{c}	$7.62 \pm 0.36^{a,b}$	7.23 ± 0.50^{c}	4.54 ± 0.20^{b}	$4.63 \pm 0.78^{a,b}$	16.41 ± 1.30^{c}
UV-B + N	19.53 ± 0.46^{b}	$7.29 \pm 0.51^{a,b}$	$7.72 \pm 0.66^{b,c}$	$5.38 \pm 0.59^{a,b}$	4.11 ± 0.09^{b}	17.22 ± 1.02^{c}
N	0.000	0.165	0.079	0.070	0.772	0.045
UV-B	0.000	0.381	0.001	0.012	0.025	0.000
N × UV-B	0.006	0.040	0.265	0.916	0.226	0.176

(Table 2). Supplemental nitrogen significantly enhanced Gs. A of plant grown at ambient UV-B was increased by supplemental nitrogen. E and Ci were not affected by supplemental nitrogen. A prominent UV-B × nitrogen interaction was observed on A.

Enhanced UV-B markedly reduced Chl *a*, Chl *b*, Chl (*a* + *b*), and Car content (Table 3). On the other hand, Chl *a*, Chl *b*, and Chl (*a* + *b*) content of plants grown at ambient UV-B were increased by supplemental nitrogen, whereas supplemental nitrogen did not influence chlorophyll pigment under enhanced UV-B. Car content was increased by supplemental nitrogen. A parallel change trend in Chl *a* and Chl *b* resulted in no significant change in Chl *a/b* ratio under enhanced UV-B or supplemental nitrogen. Significant interactive effects of UV-B and nitrogen were also detected on Chl *a* content ($P = 0.001$).

The effects of enhanced UV-B and nitrogen supply on reactive oxygen species and MDA content

Enhanced UV-B did not significantly affect H_2O_2 content and the rate of O_2^- production in *P. asperata* seedlings grown without supplemental nitrogen (Fig. 1A and B), whereas H_2O_2 content and the rate of O_2^- production of plants grown with supplemental nitrogen significantly increased by enhanced UV-B. On the other hand, supplemental nitrogen significantly reduced H_2O_2 content and the rate of O_2^- production under ambient UV-B, whereas there was no statistical significance under enhanced UV-B. The interaction of UV-B × nitrogen also significantly affected H_2O_2 content ($P = 0.001$).

MDA content significantly increased by enhanced UV-B (Fig. 1C). Supplemental nitrogen significantly reduced MDA content only under ambient UV-B. Significant interaction between UV-B and nitrogen was detected on MDA content ($P < 0.001$).

The effects of enhanced UV-B and nitrogen supply on UV-B absorbing compounds and proline content

Plants grown at environmental stress could activate the defense systems themselves. Both enhanced UV-B and supplemental nitrogen significantly increased UV-B absorbing compounds content and proline content, respectively (Fig. 2). The effect of UV-B × nitrogen interaction significantly affected proline content ($P = 0.006$).

Table 2. The effects of enhanced UV-B and supplemental nitrogen on photosynthesis of *P. asperata*. Values are the mean ± SE of six replicates in column rows 1–4 and the values in the same column with different letters are significantly different from each other ($P < 0.05$). Significant effects of the two factors as well as of the interaction are indicated in column rows 5–7.

Treatment	A (μmol m^{-2} s^{-1})	E (mmol m^{-2} s^{-1})	Gs (mmol m^{-2} s^{-1})	Ci (μmol mol^{-1})
C	7.51 ± 0.06^{b}	1.58 ± 0.15^{a}	51.86 ± 3.97^{b}	$203.48 \pm 5.57^{b,c}$
N	8.86 ± 0.24^{a}	1.79 ± 0.27^{a}	84.60 ± 6.50^{a}	195.86 ± 5.71^{c}
UV-B	6.05 ± 0.50^{c}	0.91 ± 0.07^{b}	29.18 ± 2.71^{c}	229.58 ± 8.04^{a}
UV-B + N	5.08 ± 0.44^{c}	0.82 ± 0.09^{b}	49.60 ± 4.61^{b}	$218.20 \pm 4.31^{a,b}$
N	0.645	0.737	0.000	0.129
UV-B	0.000	0.000	0.000	0.001
N × UV-B	0.020	0.373	0.205	0.785

Table 3. The effects of enhanced UV-B and supplemental nitrogen on photosynthetic pigment of *P. asperata*. Values are the mean ± SE of six replicates in column rows 1–4 and the values in the same column with different letters are significantly different from each other ($P < 0.05$). Significant effects of the two factors as well as of the interaction are indicated in column rows 5–7.

Treatment	Chl *a* (mg g^{-1} FW)	Chl *b* (mg g^{-1} FW)	Chl *a/b*	Chl (*a* + *b*) (mg g^{-1} FW)	Car (mg g^{-1} FW)
C	0.35 ± 0.01^b	0.10 ± 0.00^b	3.50 ± 0.38^a	0.45 ± 0.02^b	0.08 ± 0.00^b
N	0.43 ± 0.00^a	0.12 ± 0.00^a	3.58 ± 0.31^a	0.55 ± 0.00^a	0.10 ± 0.00^a
UV-B	0.28 ± 0.01^c	0.07 ± 0.01^c	4.00 ± 0.49^a	0.35 ± 0.05^c	0.04 ± 0.00^d
UV-B + N	0.30 ± 0.00^c	0.08 ± 0.01^c	3.75 ± 0.24^a	0.37 ± 0.00^c	0.06 ± 0.00^c
N	0.000	0.041	0.132	0.040	0.001
UV-B	0.000	0.000	0.085	0.000	0.000
N × UV-B	0.001	0.405	0.320	0.124	0.531

The effects of enhanced UV-B and nitrogen supply on the activity of antioxidant enzymes

Enhanced UV-B changed the activity of antioxidant enzymes (Fig. 3). POD, SOD, and GR activity significantly increased by enhanced UV-B, whereas APX and CAT activity in plant grown without supplemental nitrogen increased by enhanced UV-B. On the other hand, supplemental nitrogen enhanced POD, APX, and CAT activity under ambient UV-B, whereas supplemental nitrogen reduced the activity of antioxidant enzymes under enhanced UV-B. Significant interaction between nitrogen and UV-B were also found on POD, SOD, APX, GR, and CAT activity ($P \leqslant 0.001$).

Discussion

Effects of enhanced UV-B on *P. asperata*

Previous studies have shown that enhanced UV-B could inhibit shoot growth and biomass production and affect biomass allocation in forest tree species (De La Rose et al. 2003, Kossuth and Biggs 1981, Sullivan and Teramura 1992). The results were similar with our study. But we also found that enhanced UV-B also significantly inhibited underground growth (root DW) of *P. asperata* seedlings (Table 1), and needle DW of plants grown at supplemental nitrogen was reduced by enhanced UV-B. The results implied that the root of *P. asperata* seedlings was more susceptible to UV-B radiation than over-ground parts, and the sensitivity of over-ground parts to high UV-B was modified by nitrogen nutrition.

The reduced growth was closely related to a strong reduction in photosynthesis under high UV-B level (Kulandaivelu et al. 1989). The negative effects of UV-B on photosynthetic processes have been demonstrated (Allen et al. 1998, Correia et al. 2005, Zeuthen et al. 1997). Correia et al. (2005) reported that enhanced UV-B reduced A, E, Gs, and increased the Ci of maize. Similar results were found in this paper. Ci can be used to

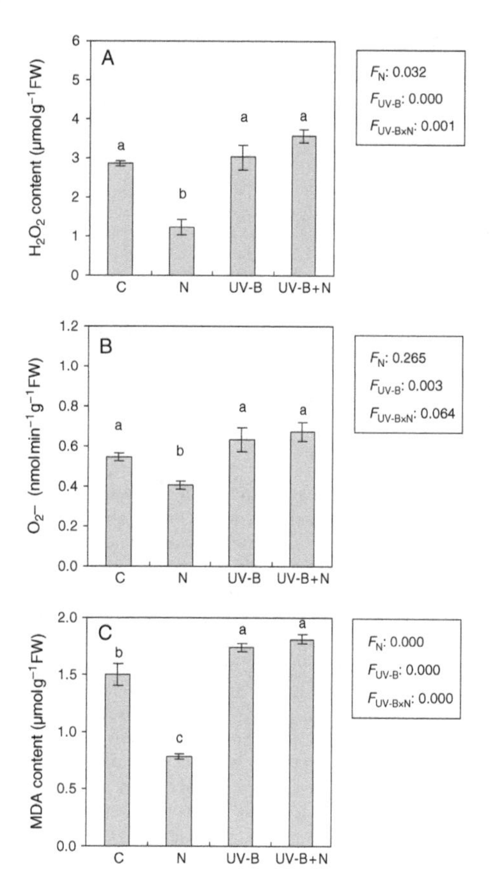

Fig. 1. (A) H_2O_2 content, (B) the rate of O_2^- production, and (C) MDA content of *P. asperata* affected by enhanced UV-B and supplemental nitrogen. The bars with different letters are significantly different from each other ($P < 0.05$). Values are means of six replicates ± SE. F_N, supplemental nitrogen effect; F_{UV-B}, UV-B radiation effect; $F_{UV-B \times N}$, interaction effect of UV-B and nitrogen.

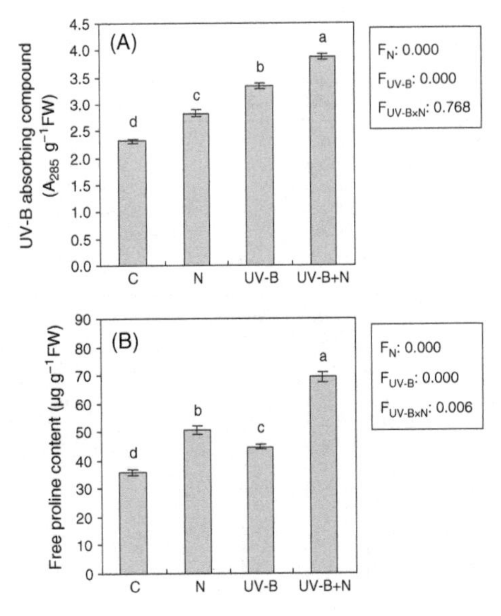

Fig. 2. (A) UV-B absorbing compoundsand (B) proline content of *P. asperata* affected by enhanced UV-B and nitrogen supply. The bars with different letters are significantly different from each other ($P < 0.05$). Values are means of six replicates ± SE. F_N, supplemental nitrogen effect; F_{UV-B}, UV-B radiation effect; $F_{UV-B \times N}$, interaction effect of UV-B and nitrogen.

discriminate between changes in A resulting from stomatal limitation or non-stomatal limitations (Farquhar and Sharkey 1982). Our results indicated that the drop in photosynthesis in high UV-B treated plants resulted from non-stomatal rather than stomatal limitations.

Decreases of Chl ($a + b$) and Car content were observed in this paper and other papers (Casati et al. 2001, Correia et al. 2005). The decreases of Chl ($a + b$) content may be due to the decreases of Car because Car protect chlorophyll from photooxidative destruction (Singh 1996). At the same time, Car content is negative related to photosynthetic efficiency (Middleton and Teramura 1993). The decrease of Car content might result from the damage of UV-B radiation or protect plant from the decrease of photosynthetic efficiency.

Enhanced UV-B radiation produces oxidative stress and increases reactive oxygen species (ROS) in plants, such as singlet oxygen, H_2O_2, and hydroxyl radicals (Mackerness et al. 1999, Santos et al. 2004). ROS affected a number of metabolic functions in plants (Carletti et al. 2003). In this study, enhanced UV-B significantly increased H_2O_2 content and the rate of O_2^- production in needles only under supplemental nitrogen, suggesting that supplemental nitrogen not only did not alleviate oxidative stress of plants suffered from high UV-B, but also aggravated the effects of UV-B stress on plants.

Enhanced UV-B with the high energy could reach the cell organelles and increase the breakdown of lipid membrane. The degree of lipid peroxidation is often assessed by MDA content and is usually used as an indicator in stress physiology of plants (Yu et al. 2004). Enhanced UV-B significantly increased MDA content in the paper, indicating the increase in lipid peroxidation. The result was in agreement with the result of other studies (Alexieva et al. 2001, Queiroz et al. 1998).

Proline metabolism is a typical metabolism of the biochemical adaptation in living organisms subjected to stress conditions (Delauney and Verma 1993). Proline could function as a hydroxyl radical scavenger to prevent membrane damage and protein denaturation (Ain-Lhout et al. 2001). Although mainly involved in the water-stress syndrome, proline reportedly accumulated in the shoots of rice, mustard, and mung bean seedlings exposed to UV radiation (Saradhi et al. 1995), which may protect plant cells against peroxidative processes (Saradhi et al. 1995). On the other hand, UV-B absorbing compounds can protect plant from the harm of UV-B radiation by reductions in the transmittance of UV photons through leaf tissue (Day and Neale 2002). In this study on *P. asperata* seedlings, enhanced UV-B radiation significantly increased proline content and UV-B absorbing compounds (Fig. 2), suggesting the important role of proline and UV-B absorbing compounds in enhanced UV-B tolerance. This may provide an ecological adaptation for young seedlings by enhanced defense substance content under stress conditions.

SOD, APX, POD, GR, and CAT are important enzymes in plant that protect plants against oxidative damage. Many deleterious effects on photosynthesis by UV-B could be caused by the generation of free radicals that destroy various components of the photosynthesis apparatus (Rajagopal et al. 2000). We found that enhanced UV-B also induced the activity of POD, GR, and SOD as the increasing requirement of screening ROS. UV-B exposure resulted in increased CAT, SOD, and APX in tomato leaves (Santos et al. 2004) and in increased SOD, APX, and PPOX in Arabidopsis thaliana (Rao et al. 1996). ROS induced antioxidative enzymes confer tolerance of stressful conditions.

Effects of nitrogen supply on *P. asperata*

In this study, total biomass increased by supplemental nitrogen only under ambient UV-B. This showed that supplemental nitrogen could alleviate the injury of

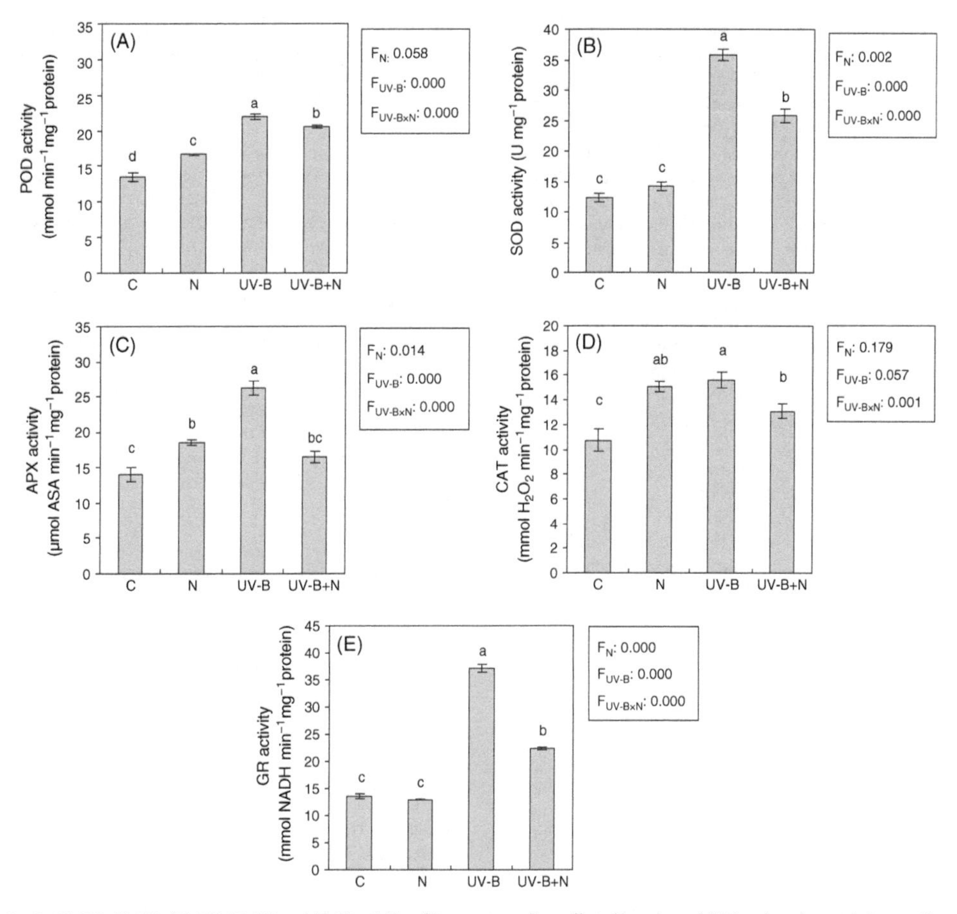

Fig. 3. (A) POD, (B) SOD, (C) APX, (D) CAT, and (E) GR activities of *P. asperata* seedlings affected by enhanced UV-B and supplemental nitrogen. The bars with different letters are significantly different from each other ($P < 0.05$). Values are means of six replicates ± SE. F_N, supplemental nitrogen effect; F_{UV-B}, UV-B radiation effect; $F_{N \times UV-B}$, interaction effect of UV-B and nitrogen.

ambient UV-B on plant biomass accumulation, whereas did not significantly ease the influence of high UV-B on plant biomass accumulation.

Nitrogen nutrition can affect the process of photosynthesis and related gas exchange (Li et al. 2004). A was stimulated by supplemental nitrogen under ambient UV-B, as in the study of Nakaji et al. (2001). In the contrast, supplemental nitrogen did not affect the sensitivity of photosynthetic rate to high UV-B in this study.

Supplemental nitrogen enhanced chlorophyll content under ambient UV-B. The results may explain the conclusion in part that A was stimulated by supplemental nitrogen. However, chlorophyll content in needles grown at enhanced UV-B were not affected by supplemental nitrogen implied that supplemental nitrogen could not ease the harmful effects of enhanced UV-B radiation on photosynthetic pigment.

MDA content in needles grown at ambient UV-B decreased by supplemental nitrogen, indicating that lipid peroxidation of cell membrane reduced, whereas MDA content was not affected by supplemental nitrogen under enhanced UV-B radiation. This showed that supplemental nitrogen did not alleviate the damage of enhanced UV-B radiation on cell membrane.

Supplemental nitrogen evidently increased proline content in needles. The result was similar to Sánchez et al. (2002). Stimulation of proline accumulation by supplemental nitrogen may indicate that it is a nitrogen-storage compound (Ahmad and Hellebust 1988), and that synthesis and accumulation of proline are simulated by supplemental nitrogen (Sánchez et al. 2002). However, Mulholland and Otte (2001) found that proline in leaves of *Spartina anglica* was not significantly affected by nitrogen treatments. Supplemental nitrogen increased UV-B absorbing compounds; this response offered supplemental protection of plant to UV-B radiation.

Supplemental nitrogen reduced H_2O_2 content and the rate of O_2^- production under ambient UV-B. The results were similar to Xiao et al. (1998) and Ramalho et al. (1998). This may be related to the increase of antioxidant enzymes activity (POD, APX, and CAT) under supplemental nitrogen (Fig. 3A, C, and D). On the other hand, nitrogen nutrition can improve light reaction and dark reaction of photosynthetic organization, and reduced deoxidization capacity, which lead to the reduced rate of ROS production and the less ROS accumulation (Xiao et al. 1998). However, supplemental nitrogen reduced antioxidant enzymes activity under enhanced UV-B radiation, suggesting that the effects of nitrogen on antioxidant enzymes were inhibited by enhanced UV-B.

Interactive effects of enhanced UV-B and nitrogen supply on *P. asperata*

Levizou and Manetas (2001) reported that supplemental UV-B radiation improved growth in *Phlomis fruticosa* at high nutrient level, whereas greater growth inhibition by UV-B has been reported in nitrate-replete than nitrate-deficient crop plants (Hunt and McNeil 1998). Tosserams et al. (2001) reported that photosynthetic rate of *Plantaago lanceolata* with high UV-B was not influenced by differential quantities of multiple mineral supply. This paper clearly showed that the effects of UV-B radiation on photosynthetic parameters and antioxidant defense parameters were dependent upon its nitrogen nutrition. At the same time, photosynthetic parameters and antioxidant defense parameters affected by nitrogen nutrition were also dependent on UV-B level. Supplemental nitrogen was favorable for the growth of plant under ambient UV-B, as could be seen by the increase of plant growth parameters, A, photosynthetic pigment, proline content, UV-B absorbing compounds and antioxidant enzymes activity, and the decrease of ROS and MDA content, whereas supplemental nitrogen makes *P. asperata* seedlings more sensitive to enhanced UV-B radiation, although supplemental nitrogen increased some antioxidant indexes under enhanced UV-B. Obviously, supplemental nitrogen could not overcome the harmful effects of high UV-B radiation on plants.

Acknowledgements – During this work, the senior author was supported by the National Natural Science Foundation of China (No. 30530630), the Talent Plan of the Chinese Academy of Sciences and Knowledge Innovation Engineering of the Chinese Academy of Sciences.

References

Ahmad I, Hellebust JA (1988) The relationship between inorganic nitrogen metabolism and proline accumulation in osmoregulatory responses of two euryhaline microalgae. Plant Physiol 88: 348–354

Ain-Lhout F, Zunzunegui M, Diaz Barradas MC, Tirado R, Clavijo A, Carcia Novo F (2001) Comparison of proline accumulation in two Mediterranean shrubs subjected to natural and experimental water deficit. Plant Soil 230: 175–183

Alexieva V, Sergiev I, Mapelli S, Karanov E (2001) The effect of drought and ultraviolet radiation on growth and stress markers in pea and wheat. Plant Cell Environ 24: 1337–1344

Allen DJ, Nogués S, Baker NR (1998) Ozone depletion and increased UV-B radiation: is there a real threat to photosynthesis? J Exp Bot 49: 1775–1788

Barnes BV, Zak DR, Denton SR, Spurr SH (1998) Forest Ecology, 4th Edn. John Wiley & Sons, Inc., New York, p 774

Bates LS, Waldren RP, Teare ID (1973) Rapid determination of free proline for water studies. Plant Soil 39: 205–208

Baumbusch LO, Eiblmeier M, Schnitzler JP, Heller W, Sandermann H, Polle A (1998) Interactive effects of ozone and low UV-B radiation on antioxidants in spruce (*Picea abies*) and pine (*Pinus sylvestris*) needles. Physiol Plant 104: 248–254

Becana M, Aparicio-Tejo P, Irigoyen JJ, Sánchez-Díaz M (1986) Some enzymes of hydrogen peroxide metabolism in leaves and root nodules of *Medicago sativa*. Plant Physiol 82: 1169–1171

Binkley D, Son Y, Valentine DW (2000) Do forests receive occult inputs of nitrogen? Ecosystems 3: 321–331

Bowden RD, Davidson E, Savage K, Arabia C, Steudler P (2004) Chronic nitrogen additions reduce total soil respiration and microbial respiration in temperate forest soils at the Harvard forest. Forest Ecol Manage 196: 43–56

Bradford MM (1976) A rapid and sensitive method for quantification of microgram quantities of protein utilizing the principle of protein-dye binding. Anal Biochem 72: 248–254

Caldwell MM (1971) Solar ultraviolet radiation and the growth and development of higher plants. In: Giese AC

(ed) Phytophysiology. Academic Press, New York, pp 131–177
Carletti P, Masi A, Wonisch A, Grill D, Tausz M, Ferretti M (2003) Changes in antioxidant and pigment pool dimensions in UV-B irradiated maize seedlings. Environ Exp Bot 50: 149–157
Casati P, Lara MV, Andreo CS (2001) Regulation of enzymes involved in C_4 photosynthesis and the antioxidant metabolism by UV-B radiation in *Egeria densa*, a submersed aquatic species. Photosynth Res 71: 251–264
Correia CM, Moutinho-Pereira JM, Coutinho JF, Björn LO, Torres-Pereira JMG (2005) Ultraviolet-B radiation and nitrogen affect the photosynthesis of maize: a Mediterranean field study. Eur J Agron 22: 337–347
Costa H, Gallego SM, Tomaro ML (2002) Effects of UV-B radiation on antioxidant defense system in sunflower cotyledons. Plant Sci 162: 939–945
Day TA, Neale PJ (2002) Effects of UV-B radiation on terrestrial and aquatic primary producers. Ann Rev Ecol Syst 33: 371–396
De La Rose TM, Julkunen-Tiitto R, Lehto T, Aphalo PJ (2001) Secondary metabolites and nutrient concentrations in silver birch seedlings under five levels of daily UV-B exposure and two relative nutrient addition rates. New Phytologist 150: 121–131
De La Rose TM, Aphalo PJ, Lehto T (2003) Effects of ultraviolet-B radiation on growth, mycorrhizas and mineral nutrition of silver birch (*Betula pendula* Roth) seedlings grown in low-nutrient conditions. Global Change Biol 9: 65–73
Delauney AJ, Verma DPS (1993) Proline biosynthesis and osmo-regulation in plants. Plant J 4: 215–223
Executive Summary (2003) Environmental effects of ozone depletion and its interactions with climate change: 2002 assessment. Photochem Photobiol Sci 2: 1–4
Farquhar GD, Sharkey TD (1982) Stomatal conductance and photosynthesis. J Ann Rev Plant Physiol 33: 317–345
Field C, Mooney HA (1986) The photosynthesis – nitrogen relationship in wild plants. In: Givnish TJ (ed) On the Economy of Plant Form and Function. Cambridge University Press, London, pp 25–55
Hunt JE, McNeil DL (1998) Nitrogen status affects UV-B sensitivity of cucumber. Aust J Plant Physiol 25: 79–86
Jansen MAK (2002) Ultraviolet-B radiation effects on plants: induction of morphogenic responses. Physiol Plant 116: 423–429
Jordan BR (1996) The effects of ultraviolet-B radiation on plants: a molecular perspective. Adv Bot Res 22: 97–162
Jordan BR (2002) Molecular response of plant cells to UV-B stress. Funct Plant Biol 29: 909–916
Ke D, Wang A, Sun G, Dong L (2002) The effect of active oxygen on the activity of ACC synthase induced by exogenous IAA. Acta Bot Sin 44: 551–556 (in Chinese)
Keski-Saari S, Julkunen-Tiitto R (2003) Resource allocation in different parts of juvenile mountain birch plants: effect of nitrogen supply on seedling phenolics and growth. Physiol Plant 118: 114–126
Kossuth SV, Biggs RH (1981) Ultraviolet-B radiation effects on early seedling growth of Pinaceae species. Can J Forest Res 11: 243–248
Kulandaivelu G, Margatham S, Nedunchezhian N (1989) On the possible control of ultraviolet-B induced response in growth and photosynthetic activities in higher plants. Physiol Plant 76: 398–404
Lavola A, Aphalo PJ, Lahti M, Julkunen-Tiitto R (2003) Nutrient availability and the effect of increasing UV-B radiation on secondary plant compounds in Scots pine. Environ Exp Bot 49: 49–60
Levizou E, Manetas Y (2001) Combined effects of enhanced UV-B radiation and additional nutrients on growth of two Mediterranean plant species. Plant Ecol 154: 181–186
Li DJ, Mo JM, Fang YT, Pen SL, Gundersen P (2003) Impact of nitrogen deposition on forest plants. Acta Ecol Sin 23: 1891–1900 (in Chinese)
Li DJ, Mo JM, Fang YT, Cai XY, Xue JH, Xu GL (2004) Effects of simulated nitrogen deposition on growth and photosynthesis of *Schima superba, Castanopsis chinensis* and *Cryptocarya concinna* seedlings. Acta Ecol Sin 24: 876–882 (in Chinese)
Lichtenthaler HK (1987) Chlorophylls and carotenoids: pigments of photosynthetic biomembranes. Methods Enzymol 148: 350–382
Liu Q (2002) Ecological Research on Subalpine Coniferous Forests in China. Sichuan University Press, Chengdu, pp 33–52 (in Chinese)
Mackerness SAH, Jordan BR, Thomas B (1999) Reactive oxygen species in the regulation of photosynthetic genes by ultraviolet-B radiation (UV-B: 280–320 nm) in green and etiolated buds of pea (*Pisum sativum* L.). J Photochem Photobiol B 48: 180–188
Magill AH, Aber JD, Hendricks JJ, Bowden RD, Melillo JM, Steudler PA (1997) Biogeochemical response of forest ecosystems to simulated chronic nitrogen deposition. Ecol Appl 7: 402–415
Middleton EM, Teramura AH (1993) The role of flavonol glycosides and carotenoids in protecting soybean from ultraviolet-B damage. Plant Physiol 103: 741–752
Mulholland MM, Otte ML (2001) The effects of nitrogen supply and salinity on DMSP, glycine betaine and proline concentrations in leaves of *Spartina anglica*. Aquat Bot 71: 63–70
Nakaji T, Fukami M, Dokiya Y, Izuta T (2001) Effects of high nitrogen load on growth, photosynthesis and nutrient status of *Cryptomeria japonica* and *Pinus desiflora* seedlings. Trees 15: 453–461
Pinto ME, Casati P, Hsu TP, Ku MS, Edwards GE (1999) Effects of UV-B radiation on growth, photosynthesis, UV-B absorbing compounds and NADP-malic enzyme in bean

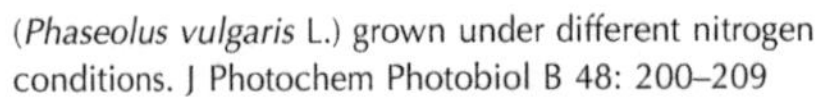

(*Phaseolus vulgaris* L.) grown under different nitrogen conditions. J Photochem Photobiol B 48: 200–209

Prochazkova D, Sairam RK, Srivastava GC, Singh DV (2001) Oxidative stress and antioxidant activity as the basis of senescence in maize leaves. Plant Sci 161: 765–771

Queiroz CGS, Alonso A, Mares-Guia M, Magalhäes AC (1998) Chilling-induced changes in membrane fluidity and antioxidant enzyme activities in *Coffea arabica* L. roots. Biol Plant 41: 403–413

Rajagopal S, Murthy SD, Mohanty P (2000) Effect of ultraviolet-B radiation on intact cells of the Cyanobacterium *Spirulina platensis*: characterization of the alterations in the thylakoid membranes. J Photochem Photobiol B 54: 61–66

Ramalho JC, Campos PS, Teixeira M, Nunes MA (1998) Nitrogen dependent changes in antioxidant system and in fatty acid composition of chloroplast membranes from *Coffea arabica* L. plants submitted to high irradiance. Plant Sci 135: 115–124

Rao MV, Paliyath G, Ormrod DP (1996) Ultraviolet-B and ozone-induced biochemical changes in antioxidant enzymes of *Arabidopsis thaliana*. Plant Physiol 110: 125–136

Sánchez E, Ruiz JM, Romero L (2002) Proline metabolism in response to nitrogen toxicity in fruit of French Bean plants. Sci Hortic 93: 225–233

Santos I, Fidalgo F, Almeida JM, Salema R (2004) Biochemical and ultrastructural changes in leaves of potato plants grown under supplementary UV-B radiation. Plant Sci 167: 925–935

Saradhi PP, Alia SA, Prasad KV, Arora S (1995) Proline accumulates in plants exposed to UV radiation and protects them against UV induced peroxidation. Biochem Biophys Res Commun 209: 1–5

Singh A (1996) Growth, physiological, and biochemical responses of three tropical legumes to enhanced UV-B radiation. Can J Bot 74: 135–139

Sullivan JH, Teramura AH (1992) The effects of ultraviolet-B radiation on loblolly pine. 2. Growth of field-grown seedlings. Trees 6: 115–120

Tosserams M, Smet J, Magendans E, Rozema J (2001) Nutrient availability influences UV-B sensitivity of *Plantago lanceolata*. Plant Ecol 154: 159–168

Vitousek PM, Howarth RW (1991) Nitrogen limitation on land and in sea. How can it occur? Biogeochemistry 13: 87–115

Xiao K, Zhang RX, Qian WP (1998) The physiological mechanism of senescence and photosynthetic function decline of flag leaf in wheat regulated by nitrogen nutrition. Plant Nutrition and Fertilizer Science 4: 371–378 (in Chinese)

Xu XL, Ou YH, Pei ZY, Zhou CP (2003) Fate of N^{15} labeled nitrate and ammonium salts added to an Alpine Meadow in the Qinghai-Xizang Plateau. Acta Bot Sin 45: 276–281 (in Chinese)

Yu J, Tang XX, Zhang PY, Tian JY, Cai HJ (2004) Effects of CO_2 enrichment on photosynthesis, lipid peroxidation and activities of antioxidative enzymes of *Platymonas subcordiformis* subjected to UV-B radiation stress. Acta Bot Sin 46: 682–690

Zeuthen J, Mikkelsen TN, Paludan-Müller G, Ro-Poulsen H (1997) Effects of increased UV-B radiation and elevated levels of tropospheric ozone on physiological processes in European beech (*Fagus sylvatica*). Physiol Plant 100: 281–290

Edited by D. Campbell